Global Resources

Global Resources

Conflict and Cooperation

Edited by

Roland Dannreuther
Professor of International Relations, University of Westminster, UK

and

Wojciech Ostrowski
Research Fellow in International Politics of Energy and Mineral Resources,
Centre for Energy, Petroleum and Mineral Law Policy, University of Dundee, UK

First published 2013 by
PALGRAVE MACMILLAN

Palgrave Macmillan in the UK is an imprint of Macmillan Publishers Limited,
registered in England, company number 785998, of Houndmills, Basingstoke,
Hampshire RG21 6XS.

Palgrave Macmillan in the US is a division of St Martin's Press LLC,
175 Fifth Avenue, New York, NY 10010.

Palgrave Macmillan is the global academic imprint of the above companies
and has companies and representatives throughout the world.

Palgrave® and Macmillan® are registered trademarks in the United States,
the United Kingdom, Europe and other countries.

ISBN 978–0–230–36050–1

This book is printed on paper suitable for recycling and made from fully
managed and sustained forest sources. Logging, pulping and manufacturing
processes are expected to conform to the environmental regulations of the
country of origin.

A catalogue record for this book is available from the British Library.

A catalog record for this book is available from the Library of Congress.

Contents

Part IV   Scarcity, Technology and Future Supply

Figures and Tables

Figures

Tables

Acknowledgements

The authors would like to acknowledge that this book is, in substantial part, an outcome of the EU Project for Natural Resources (*Polinares*), which was funded through the European Commission's 7th Framework Programme (project number 224156). Particular thanks go to Domenico Rossetti di Valdalbero, Principal Administrator for DG Research at the European Commission, and Philip Andrews-Speed, the scientific coordinator for the project.

The authors and publishers are grateful to the following for permission to reproduce copyright material:

Goldman Sachs for Exhibit 4, from Jeffery Currie et al., *The Revenge of the Old 'Political' Economy*, Goldman Sachs, 14 March 2008.

Abbreviations

ANC	African National Congress
ARAMCO	Arabian American Oil Company
ASPO	Association for the Study of Peak Oil and Gas
BHP	Broken Hill Proprietary
CCGT	Combined Cycle Gas Turbine
CERA	Cambridge Energy Research Associates
CFP	Compagnie Française des Pétroles
CIC	China Investment Corporation
CNMC	China Nonferrous Metals Corporation
CNOOC	China National Offshore Oil Corporation
CRS	Corporate Social Responsibility
CVRD	Companhia Vale do Rio Doce
ECT	European Charter Treaty
EIA	Energy Information Administration (US)
EITI	Extractive Industries Transparency Initiative
EOR	Enhanced Oil Recovery
EPA	Environmental Protection Agency (US)
ERCB	Energy Resources Conservation Board (Canada)
FIRB	Foreign Investment Review Board (Australia)
FRELIMO	Liberation Front of Mozambique
FSU	Former Soviet Union
GCC	Gulf Cooperation Council
GHG	Greenhouse Gas
IEA	International Energy Agency
IEF	International Energy Forum
IFPEN	IFP Energies Nouvelles
IIED	International Institute for Environment and Development
IMF	International Monetary Fund
IOCs	International Oil Companies
IPC	Iraq Petroleum Company
IPCC	Intergovernmental Panel on Climate Change
IPO	Initial Public Offering
ISI	Import Substitution Industrialisation
JOA	Joint-Operating Agreements
JODI	Joint Oil Data Initiative
JV	Joint Venture
KIO	Kuwait Investment Office
KOC	Kuwait Oil Company

LKAB	Luossavaara-Kiirunavaara Aktiebolag
LNG	Liquefied Natural Gas
LPG	Liquefied Petroleum Gas
MIM	Mount Isa Mines
MMSD	Mining, Minerals and Sustainable Development
MNC	Multinational Corporation
MPLA	People's Movement for the Liberation of Angola
NGLs	Natural Gas Liquids
NGOs	Non-Governmental Organisations
NIEO	New International Economic Order
NMC	National Mining Company
NOCs	National Oil Companies
NRC	National Resource Company
NYMEX	New York Metal Exchange
OBT	Obsolescing Bargaining Theory
OECD	Organisation for Economic Cooperation and Development
OPEC	Organization of Petroleum Exporting Countries
PDVSA	Petróleos de Venezuela, S.A.
PSA	Production Sharing Agreement
PSC	Production Sharing Contract
PWYP	Publish What You Pay
RST	Rentier State Theory
SA	Service Agreement
SAP	Structural Adjustment Programme
SOE	State-Owned Enterprise
SWF	Sovereign Wealth Found
TNC	Transitional Corporation
TPA	Third Party Access
UMHK	Union Minière du Haut Katanga
UNCTAD	United Nations Conference on Trade and Development
UNDP	United Nations Development Programme
UNEP	United Nations Environment Programme
UNITA	National Union for the Total Independence of Angola
UNOCAL	Union Oil Company of California
URR	Ultimate Recoverable Resources
USGS	United State Geological Survey
WBMS	World Bureau of Metal Statistics
WTI	West Texas Intermediate
WTO	World Trade Organization

Contributors

Ariel Bergmann is an energy economist at the Centre of Energy, Petroleum and Mineral Law and Policy (CEPMLP), a multidisciplinary institute within the University of Dundee, United Kingdom. His research is oriented towards renewable and alternative energy issues, specifically social value and willingness to pay for alternative energy sources. He is also interested in the interface between energy projects and environmental impacts at the local or regional level.

Michael Bradshaw is Professor of Human Geography and former Head in the Department of Geography at the University of Leicester, UK. His PhD is from the University of British Columbia, Canada. His research is on resource geography with a particular focus on the economic geography of Russia and global energy security. He is Honorary Senior Research Fellow at the Centre for Russian and East European Studies in the University of Birmingham and Visiting Senior Research Fellow at the Oxford Institute for Energy Studies. He is currently conducting research project on the geopolitical economy of global gas security and governance, funded by the UK Energy Research Centre. He has a forthcoming book entitled *Global Energy Dilemmas*.

Murtala Chindo was educated at UDU Sokoto, Imperial College London, and has a doctorate degree in Geography from the University of Leicester. He is Honorary Research Associate, University of Leicester, and Lecturer in the Department of Geography at IBB University, Lapai, Nigeria. From 2004 to 2007, he was a policy researcher on natural resources and infrastructure for a presidential think tank that provided policy advice to the president of Nigeria. His research interests include communities, resource extraction and development, resource governance, mining and geographic information systems.

Chris Cragg is a freelance journalist, specializing in energy and environmental issues. He was editor of the *Financial Times Energy Economist* for 14 years and worked as a writer-researcher at BP for 4 years. He writes regularly for the European Energy Review and for Platts. He is currently engaged in advising the Nigerian government on pollution issues.

Patrick Criqui is a senior researcher at CNRS and director of the Economics of Sustainable Development and Energy Laboratory (EDDEN, UPMF-CNRS Grenoble). His research initially focused on the economics of solar energy and modelling of international energy markets. He then developed global long-term energy model, POLES, which is currently being used by the European Commission and by different administrations and companies in Europe to analyse the economics of climate policies. He was a lead author in IPCC's Working Group 3 (Nobel Peace Prize in 2007). He is member of the Economic Council for Sustainable Development headed by the French Minister for Ecology and member of the scientific council of the Local Climate Action Plan of the Grenoble metropolitan area.

Roland Dannreuther is Professor of International Relations and Head of the Department of Politics and International Relations, University of Westminster. He is also International Fellow at the Department of International Relations, Tbilisi State University, Georgia. His research revolves around the areas of security studies, energy politics and international relations with a regional focus on Russia, Central Asia and the Middle East. His current research is on energy security and international energy politics, with a particular focus on China's international energy strategy and the future dynamics of international energy politics. Recent publications include *China, Oil and Global Politics* (2011) co-authored with Philip Andrews-Speed.

Evelyn Dietsche is a development economist and public policy analyst who has held positions in the public and private sectors and academia. She has recently completed a position with a global energy company where she was instrumental in developing the company's sustainable development and broader licence-to-operate strategy. She is now a principal consultant with Oxford Policy Management and an honorary lecturer with the Centre for the Energy, Petroleum and Mineral Law and Policy at the University of Dundee.

Magnus Ericsson is co-founder and chairman of the Raw Materials Group (RMG) and professor of mineral economics at the Luleå University of Technology in Sweden. His interest in the social, political and economic impacts of mining goes back to the 1970s. He has a special focus on African mining.

David Humphreys is an independent consultant based in London. He was previously chief economist of Rio Tinto and Norilsk Nickel. David has written and lectured extensively on the economics of the mining industry. Recent publications include 'Transatlantic Mining Corporations in the Age of Resource Nationalism', *Transatlantic Academy* (2012). David is an honorary lecturer at the University of Dundee and a non-executive director of Russian gold miner Petropavlovsk.

Giacomo Luciani is the Scientific Advisor of the Master's in International Energy at the Paris School of International Affairs of Sciences-Po in Paris; adjunct professor of International Affairs and co-director of the Executive Master in International Oil and Gas Leadership at the Graduate Institute of

International Affairs in Geneva; and a Princeton University Global Scholar affiliated to the Near Eastern Studies Department and the Woodrow Wilson School of Public and International Affairs. His research interests include the political economy of the Middle East and North Africa and the geopolitics of energy. His work has focused primarily on the economic and political dynamics of *rentier* states and issues of development in the GCC countries.

Wojciech Ostrowski is Post-Doctoral Research Fellow in International Politics of Energy and Mineral Resources at the Centre for Energy, Petroleum and Mineral Law and Policy, University of Dundee, and Visiting Lecturer at the Department of Politics and International Relations, University of Westminster. He is the author of *Politics and Oil in Kazakhstan* (2010).

Patrik Söderholm is Professor of Economics at Luleå University of Technology, Sweden. His research is focused on the economics of energy, natural resources and the environment. Söderholm has been a guest researcher at the Massachusetts Institute of Technology (MIT) and the International Institute for Applied Systems Analysis (IIASA), Vienna.

Paul Stevens is Senior Research Fellow at Chatham House. He is also Professor Emeritus at the University of Dundee and Visiting Professor at University College London. He has published extensively on energy economics, the international petroleum industry, economic development issues and the political economy of the Gulf. In March 2009, he was presented with the OPEC Award in recognition of his outstanding work in the field of oil and energy research.

Introduction: The Dynamics of Conflict and Cooperation

Roland Dannreuther

Oil, gas and minerals are vital natural resources that meet crucial human needs. Whether for transportation, for heating, or for everyday goods and services, these resources constitute essential raw material inputs. Modern civilization would struggle to survive without readily available access to these resources at reasonable and affordable prices. It is for this reason that oil, gas and a large number of other minerals are considered to be 'strategic' resources, critical for national and global well-being and prosperity. The fact that, generally speaking, there are well-functioning and sophisticated markets for these resources, and that they are usually available at prices which are affordable (at least for the developed world), reflects the fact that cooperation exists between a variety of actors – states, companies and communities – and at local, national, regional and international levels. However, there are also inevitably sources of conflict, whether these be unresolved historical legacies, disputes over pricing or relations of distrust between the various actors, such as between companies and states or between importing and exporting states. In practice, both cooperation and conflict are an ineradicable part of the complex life cycle of these vital resources from physical extraction to their final use and consumption.

Global Resources provides critical insights into the dynamic processes of conflict and cooperation in relation to oil, gas and minerals. The book recognizes that there is no easy and clear separation between areas of cooperation and of conflict. The reality is that relations of conflict and cooperation co-exist in complex and continually evolving ways. This is particularly evident in the histories of the industries for oil, gas and minerals, which is the focus of the first part of the book. As a summary of the key historical periods or 'regimes' identified through the historical chapters, there was an agreement that at least five key periods can arguably be identified:

- 1840–1914 Imperial Liberalism
- 1914–45 Mercantilism and War Economy
- 1945–80 State Interventionism and Socialism

- 1980–2008 Liberal Capitalism
- 2008– State Capitalism

What becomes clear is that different periods of history have distinctive forms and balances of conflict and cooperation. For example, the liberal imperial age of European dominance had a particular set of sources of conflict and cooperation, where the rivalry was more between Western states and companies than between North and South; this shifted dramatically after the Second World War when the onset of the Cold War and the dismantlement of the European empires led to much stronger sources of conflict between southern producing states and northern importing states.

Drawing from such historical analyses, a key issue debated in *Global Resources* is whether we are currently experiencing a similar shift in the balance of conflict and cooperation. There is a general consensus among the contributors that the 1980s and 1990s represented a period when liberal ideology and the belief in the efficiency of the market, and the associated belief in the limited role for the state in economic affairs, deeply influenced international policy towards the oil, gas and mineral markets. There is also a consensus among the contributors that during the 2000s this liberal dominance has been severely challenged and questioned. The financial crisis from 2007 onwards undermined the credibility of the neo-liberal model of financial deregulation and privatization. During the same period, the state has been more active and interventionist in the oil, gas and mineral markets, whether this be due to the increased activity of the national resource companies of many of the exporting states or to the rapidly expanding activities of the state-owned oil and mining companies of emerging countries, such as China and India. Although there is no clear agreement or consensus on how this new era might be characterized, or even whether it does represent such a critical shift, all the contributors to the book have been asked to address the implications of this apparent shift away from a more liberal capitalist to a more state interventionist or 'state capitalist' global regime.

The original inspiration of *Global Resources* came from the collaboration of a number of the contributors to the book in the EU Policy for Natural Resources (*Polinares*) project, which was a three-year (2009–12) EU-funded Framework 7 project assessing the global challenges in relation to access to oil, gas and mineral resources over the next 30 year.[1] The nature and focus of this pan-European project contributed to two distinctive and innovative features of this book. The first is the comparative dimension where the project brought together scholars and analysts from both the oil and gas and the minerals sectors, who generally only infrequently engage directly with one another. It is rare that the sources of conflict and cooperation between these different resources – oil, gas and 'hard' minerals – are compared and evaluated and this has provided a number of new and unexpected insights and perspectives. The second innovative dimension is the interdisciplinary mix, which was again very much driven by the range of disciplines represented in the *Polinares* project, extending from the hard sciences, such as geology, to the softer social sciences, such as economics and political science. The project also brought together a mix of university academics and policy-oriented analysts, all from a number of different European countries. Overall, a key intellectual benefit of the project, which has fed into this book, is as an open forum to discuss critical debates over natural resources from a truly multi-resource and interdisciplinary perspective.

Contents of the book

Global Resources has four main parts. Part I provides an historical perspective of the different resources, which are the focus of the book. In Chapter 1, Paul Stevens provides a history of the international oil industry which highlights the dramatic shift in the strategic balance of power in the 1970s when the international oil companies (IOCs), who had previously dominated the industry, ceded their primacy to the producer states in the Middle East and elsewhere. This shift in the nature of cooperation and conflict reflects, according to Stevens, the cyclical nature of the oil industry, which is driven by three different cycles – a political, a resource nationalist and an obsolescing bargaining cycle. From 2000 onwards, he sees the 'cycles beginning to swing again', with a revival of resource nationalism, state intervention and tighter markets.

In Chapter 2, David Humphreys provides a similar survey of the historical evolution of the minerals industry and identifies a number of parallels with the history of the oil industry. However, there are differences, the most notable being that there was never the same shift of power from the global mining companies to the producer states during the period of decolonization as was the case for the oil industry. Humphreys also highlights the greater salience that community conflicts and environmental concerns have had in the industry, particularly since the 1980s. However, he does note that the global mining companies are currently facing the same sorts of pressures as the IOCs, with a revival of resource nationalism in many parts of the developing world along with increased competition from state-controlled mining companies from China and other emerging countries. In Chapter 3, the final chapter of this part, a history of gas is given by Chris Cragg and he highlights the very different historical trajectory of this resource compared to both oil and the 'hard' minerals. This is due to two main factors, according to Cragg. First, it is only relatively recently that gas has been an internationally traded rather than just a local fuel. Second, the enormous infrastructural requirements for the extraction and distribution of natural gas have meant that states have always been heavily involved and the mutual dependence that this has fostered has tended to promote cooperation rather than conflict.

Part II of *Global Resources* highlights the different conceptual and theoretical frameworks, which have also had a profound impact on the dynamics of conflict and cooperation in the oil, gas and mineral sectors. In Chapter 4, Roland Dannreuther argues that the three principal theoretical traditions within international relations – realism, liberalism and radicalism – reflect different ways of conceptualizing the prospects of conflict and cooperation. Realism focuses on the prospect of resources contributing to interstate conflict; liberalism highlights the role of markets in promoting cooperation and how state intervention tends to undermine this cooperation, contributing to conflict; and the radical tradition notes how it is the legacies of the unjust and inequitable distribution of political and economic power that are the principal underlying source of conflict over natural resources. These theoretical perspectives capture much of the conflicting perceptions and views about the dynamics of conflict and cooperation in the oil, gas and minerals sectors. In Chapter 5, Wojciech Ostrowski surveys the various theories in the field of political economy that have directly addressed the issue of natural resources and development. Again, he notes that the political economy tradition has had differing perspectives at different periods – at times, natural resources such as oil, gas and minerals have been viewed as an essential aid for fast development; at other times, and particularly in the postcolonial period, such resources have been seen more negatively, even as a 'curse' and an obstacle to development. Ostrowski critiques some of these more pessimistic analyses, such as the resource curse theory, and argues that there has been too much focus on the resource-rich developing states and not enough on the role of companies and other actors in the political economy of resources.

Part III seeks to fill the gap identified by Ostrowksi and provides different perspectives from different actors involved in the oil, gas and mineral industries. In Chapter 6, Giacomo Luciani provides a number of critical insights into the relationship between companies and states in the international oil industry. This chapter can fruitfully be read alongside the history of oil provided by Stevens in Chapter 1. Luciani argues that the closest that the global oil and gas industry came to constituting a stable cooperative regime was the Sisters' Regime up to the 1960s. Retrospectively, it can be seen that, though this regime was structurally inequitable, it did provide the essential conditions for the West's post-Second World War economic miracle. Since the collapse of that regime, no effective and stable new regime has emerged and rather there has been a general weakening of the power of both companies and states, which has fuelled price volatility, underinvestment, and conflict. For Luciani, the liberal market fundamentalism of the 1980s and 1990s failed to provide the necessary conditions for stability, with an inevitable return of the state in the oil and gas industries in the more volatile conditions of the 2000s.

Chapter 7 provides a very different perspective from the level of the 'local community' rather than the transnational company. Chindo Murtala and Michael Bradshaw's contribution provides a salutary reminder that mining normally has immediate and localized affects, which can both advance as well as damage the interests of local communities. Concern for the well-being of local communities has only been a relatively recent development in the mining industry. The authors provide a case study of the challenges of including the interests of such communities in the prospective development of oil sands in Nigeria and how to avoid the example of the tragic conflict-ridden development of the Niger Delta's oil resources. In Chapter 8, Evelyn Dietsche provides an in-depth survey of the legal frameworks and contracts that determine relations between states and companies and communities. Her principal argument is that a core source of conflict over mining is located in the allocation of property rights over the resources concerned and that the liberal theory of property tights, which was dominant in the policy prescriptions promoted during the 1980s and 1990s, fails to accord sufficient attention to the obligations of the state for ensuring the public benefits of such development, including the rights of local communities. As such, and following the analysis of the other two chapters of this part, the oil, gas and minerals industries are not just governed by apolitical and self-regulating markets but are deeply intertwined with social and political concerns.

Part IV provides an opportunity to look forward and engages, in various ways, with the widespread and popular argument that natural resources like oil, gas and other minerals are scarce commodities and becoming ever scarcer. In Chapter 9, Patrick Criqui engages directly with the 'peak oil' theory, which claims that the world is on the cusp of a global energy crisis as oil production reaches its limit and begins rapidly to decline. Criqui takes an 'agnostic' position between the peak oil 'pessimists' and the anti-peak oil 'optimists', arguing that increasing demand for oil in Asia and other fast-growing regions is likely to lead to a peak of conventional oil production, if not immediately at least over the next couple of decades, but that technological advances offer opportunities to offset this through expansion of the non-conventional oil resource base. However, such a development would only lead from 'Charybdis to Scylla' as the extraction and use of non-conventional oil is environmentally damaging and will only increase greenhouse gas emissions and undermine efforts to deal with the challenge of climate change. In Chapter 10, Ariel Bergmann addresses the application of peak oil theory to gas and, like Criqui, assesses the complex linkages between technological innovation, resource scarcity and environmental concerns. He also addresses the strategic implications of the major discoveries of non-conventional gas in the US and how this 'unconventional revolution' might impact on the global gas market. In Chapter 11, the mineral peak debate is addressed by Magnus Ericcson and Patrick Soderholm and they come to a similar conclusion as the previous two authors, arguing that the key weakness of peak theories are that they ignore the institutional, environmental and technological factors which both cause and tend, over time, to resolve peaks in production. In addition, they argue that there is a real danger that, if the 'peak' debate gains a popular as well as academic traction, it could have a damaging impact on developing countries in discouraging them from using their natural resources to ensure the development of their countries.

Identifying the drivers of cooperation and conflict

It is impossible to draw a clear single or unifying conclusion from these diverse contributions to understanding the dynamics of conflict and cooperation. As is apparent in the text, there are critical differences between oil, gas and minerals, which could suggest that one is trying to compare 'apples and oranges'. There are also major differences between different regions, such as between the Middle East, Africa, the former Soviet Union and Latin America. The histories also highlight the ways in which relations of conflict and cooperation change over time. There are also different theoretical and conceptual frameworks, which significantly change the lenses through which conflict and cooperation in the oil, gas and minerals industry are understood. It also matters which actor is involved, whether it be an IOC or the NOC, a resource-producing state or an importing state, a developed or a postcolonial developing state, a host state or a local community. Each of these different actors has their own particular perspective or viewpoint.

Despite this diversity, there are nevertheless certain elements or relations which regularly recur in the analyses of *Global Resources* and which identify some of the key drivers of conflict and cooperation. The first is the most basic and intuitive and it is the relative value of the resource. This is normally identified by the price and the implicit assumption is that, broadly speaking, when prices rise for one or more of these resources then the strategic interest in such resources increases, which potentially leads to greater conflict. For example, the rise of resource nationalism and greater assertiveness of resource-producing states in the 2000s is linked in most analyses with the rapid rise in commodity prices during the same period. But it is not just price that is critical. The concept of rent is a key explanatory variable frequently cited for explaining the peculiar dynamics of the oil, gas and mineral sectors. As Humphreys and Cragg in Chapters 2 and 3 note, it is the relatively high differential rent accruing to oil which makes this resource more geopolitically significant than gas and other minerals, which have less capacity to generate rent.

A second driver for conflict and cooperation is geographical location. As Luciani notes in Chapter 6, the first time that states became deeply involved in the international oil industry was in the Middle East when the ailing Ottoman Empire was the focus of geopolitical manoeuvring between the European powers. Instability in the Middle East has constantly been a major source of concern for the oil and gas industry. Instability in Africa has similarly been a serious source of concern for importing states to ensure access to the continent's oil, gas and minerals. However, such anxieties are not apparent when such resources are found in developed Western states, such as Canada, Australia or the US. But, conversely, such fears are evident for countries which are perceived to be geopolitical challengers to the West. During the Cold War, it was the Soviet Union which was seen to be a major threat to the global oil, gas and mineral markets. Even now it is in relation to Russia, and not North Africa, that Europeans fret about their gas supplies. In the post-Cold War period, China has tended to replace the Soviet Union as the main potential threat to the global supply of oil, gas and minerals.

Geography matters, therefore, but closely connected to this is the importance of geopolitics, which can be considered a third driver. The perceptions about which country or countries are hegemonic in the international system, and which country or countries are challenging that hegemony, define the broader context of conflict and cooperation. Much of the modern history of oil, gas and minerals has been about former colonial states seeking to regain control over their resources. The post-Cold War period has similarly been a struggle between the West asserting its dominance and hegemony and the challenge to that hegemony from the emerging powers of Asia, most notably China and India. The challenge is in the form of not only a physical shift in power relations but also a shift in ideology and social and political thought. One of the central themes in the book is the implications for the oil, gas and mineral industries of the shift from a world dominated by neo-liberal ideology, the so-called Washington Consensus, to one where that ideology is challenged by a more statist and interventionist conceptualization of capitalism. Such geopolitical-driven ideological shifts have a profound impact on the ways in which the oil, gas and mineral markets are constituted and the relations of conflict and cooperation.

The fourth and fifth drivers identify key sets of relations between the various actors in the oil, gas and minerals industries, which are identified in various parts of *Global Resources* as central to the prospects for cooperation and conflict. First, there are relations between states and companies. As Stevens argues in Chapter 1, the greater part of 'conflictual or collaborative' relations boils down to the nature of the agreement 'between the owner ... and the operator' and where the owner is almost exclusively the state. At the heart of the much of the dynamics of conflict and cooperation is the struggle of the owners of these natural resources seeking to maximize the extraction or rent, while also needing the technical expertise of the companies to efficiently develop and market these resources, who are in turn seeking to maximize their profits. But as Dietsche notes in Chapter 7, it is not just the relations between states and companies which are critical; there are also the conflicts that emerge in the distribution of the benefits of the resources to the various stakeholders within the state. A key challenge, which is central to much of the academic debates in political economy as set out by Ostrowski in Chapter 5, is about ensuring that the benefits from these resources contribute to the broader development of society and are not captured by a small and unrepresentative elite. The fifth critical driver is therefore the nature of state–society relations and the degree to which the resource-rich state is a synergistic developmental state or a repressive predatory state.

The final two factors are ones that are particularly identified in the last part of the book. This is first the role of technology in potentially transforming the landscape of conflict and cooperation in the oil, gas and minerals sectors. All of the authors dealing with the 'peak' theories highlight the tendency radically to underestimate the transformative role that technology, or more simply human ingenuity, can have on future supplies. This is most dramatically seen in the US, where the technological advances in extracting unconventional gas and oil could potentially lead to an energy independence only dreamed about earlier. The consequences of this for future conflict and cooperation over global energy markets have yet to be fully assessed or understood. However, human ingenuity and technological advances not only mean that supplies of resources can be made more plentiful; they can also lead to the use of different resources to replace or substitute for existing resources.

This brings us to the final factor, which is the environment, and, as Criqui points out in Chapter 9, some of the most critical energy challenges in the future would be resolved if globally climate change mitigation measures were taken seriously and there was a genuine implementation of decarbonization policies. This broader existential challenge represented by climate change is not directly addressed in this book but it casts a critical shadow over the book. The environment also plays other important roles. As Humphreys in Chapter 2 and Murtala and Bradshaw highlight in Chapter 7, the environmental consequences of mining are a major factor affecting local communities and is something to which states and companies have had to accord far greater attention, not least due to the pressure of increasingly influential environmental NGOs. Thus, at different levels all the way from the local to the global, environmental factors do play a critical role in affecting the oil, gas and minerals sectors and in defining the context of cooperation and conflict.

These seven factors or drivers of conflict and cooperation in the oil, gas and mineral industries are far from exhaustive or inclusive. They are put forward as hopefully useful landmarks for the reader to help navigate through the rich and compelling detail and arguments of the different contributions of the book. Some of these elements will, though, be picked up in the conclusion, which will seek to draw together a number of themes in the book and provide pointers to future research.

Note

1. For details of the project, see www.polinares.eu.

Part I
Historical Legacies

1
History of the International Oil Industry

Paul Stevens

The purpose of this chapter is threefold. The first is to provide a narrative of events relating to the oil industry, both at local, regional and international level. To achieve this, six historical periods have been identified largely defined by oil market events. These set the scene for the narrative that runs throughout the chapter derived from the three cycles, described below, which have dominated the history of the international oil industry. It cannot be stressed enough how important it is that the history is understood if the present and future is to be comprehensible.[1] The second is, in each of these periods, to identify a spectrum of relations between the various players (governments – producing and consuming; companies – Majors, independents and national) ranging from conflict to tension to competition to collaboration. Finally, an attempt has been made to identify the various major transmission mechanisms between these periods. Overall, this links into the overarching theme of the book that concerns the role of state capitalism and how it influences the behaviour and motivation of the various players.

The underlying theme of the chapter is that much of these relations, be they conflictual or collaborative, derive from the agreement between the owner of the (potential) oil-in-place (outside of the US, the state) and the operator. In turn, these relationships influence the way in which the various players, that is, governments and various types of companies interact between themselves. Such relations occur and develop in the context of three cycles which characterize the industry (Stevens 2008a): the political cycle which refers to the attitude to state intervention in the allocation of economic resources in an economy; the resource nationalist cycle which relates to the attitude to allowing foreign company involvement in the sector; and finally, the obsolescing bargain cycle which concerns the willingness and ability of the state to revise the fiscal terms inherent in the agreement. A common theme that links these three cycles and runs throughout this book concerns the role of state capitalism and the nature of the regimes.

Birth of the industry: The nineteenth century

Oil in the nineteenth century was dominated by what was happening in the US (Sampson 1975; Yergin 1991). Conventionally, the history of the industry begins in 1859 when Edwin Drake drilled the first well in Titusville, Pennsylvania.[2] The early days of the industry in the US were dominated by intense competition and rivalry. This arose in large part because of the problem of property rights where the commodity in question (i.e. oil) flows in three-dimensional space and is therefore no respecter of land boundaries (Bradley 1996). Thus the Law of Capture introduced to solve this property rights problem encouraged producers to produce from their own pools as quickly as possible before their neighbours drained the reserves. The result was extreme price volatility as larger finds disrupted markets and also caused serious damage to the recovery factor of the fields. This intense rivalry was greatly aggravated by 'robber baron corporations' in what was an incredibly corrupt political environment that allowed them to behave without restraint.

Gradually however, the Standard Oil Trust of J. D. Rockefeller between 1873 and 1882 effectively consolidated the industry suppressing competition. This was driven predominantly by control over pipelines and refining. It was in effect a collaborative solution but one imposed invariably using methods that could hardly be described as pristine and often involving violence. By 1904, the Trust controlled over 85 per cent of the throughput of US refineries. This whole system was effectively destroyed in 1911 when the US Supreme Court, invoking the Sherman Anti Trust Laws of 1904, broke up the Standard Oil Trust creating a large number of what were supposed to be competing companies. However, some stability was restored to crude supply with the introduction of pro-rationing by the federal government, operating at a state level, which was intended to limit production to prevent field damage and to limit price volatility from oversupply. In effect, as will be developed below, this was the precursor to Organisation of Petroleum Exporting Countries (OPEC).

In this period the US dominated the international markets for oil. In the 1860s over half of US kerosene production was exported and by the 1880s oil products represented the fourth largest US export (Penrose 1968). As for the rest of the world in this period, other countries were beginning to produce oil. The main players in Europe were the Nobel Brothers and a group of private Russian companies. However, oil markets were largely a localized regional phenomenon given the relatively high cost of transporting oil. Thus, there was relatively little competition within these regional markets and many were effectively monopoly markets. However, oil consumption was low and dominated by the use of kerosene for lighting purposes, which had effectively pushed whale oil out of the primary energy mix. For example, in the UK, arguably the most industrialized of the countries in 1900, less than 1 per cent of primary energy consumption was petroleum (Fouquet and Pearson 1998).

The rise of oil: 1901 to the Second World War

In 1901, the Persian government awarded an oil concession to William D'Arcy. The subsequent discovery of oil at Masjid-i-Sulaiman in 1908 effectively signalled the start of a new era whose main characteristic was the availability of cheap oil outside of the US, much of it from the Middle East (Stocking 1970). A peculiar mixture of conflict and collaboration characterized this era.

The main source of conflict was the battle to secure exploration acreage in a world where the industry was dominated by eight international companies: the Majors[3] – BP (British), Shell (Anglo-Dutch), CFP (French), and Chevron, Exxon, Gulf, Mobil and Texaco (American). The story revolved around their machinations and those of their home governments – Britain, France and the US (Keating 2006). This conflict was aggravated by several factors. The First World War had identified the strategic importance of oil to fuel the machines of war.[4] In addition, oil consumption began to rise, a situation given a major boost by the development of the mass-produced car – the first Model T Ford appeared in 1912. By 1925, in the US, liquid fuels accounted for 19 per cent of total primary energy consumption (Darmstadter et al. 1971). Another factor aggravating conflict was the fact that the aftermath of the Treaty of Versailles and the creation of the League of Nations gave Britain and France very significant mandate powers over the oil-producing regions of the Middle East. The result was fierce competition by these governments (and increasingly the US) to secure concessions for their oil companies (Keating 2006).[5] The result of these interactions was the creation of the major concessions that dominated the international industry until the early 1970s.[6] The ultimate winners were reflected in the nationalities of the Majors – US, Britain and France. Prior to the First World War, Germany also played a part in this scramble for resources but subsequently was effectively pushed out of the game. Some have indeed argued that the Second World War was in part an attempt by Germany to secure access to oil reserves faced with being frozen out of the process after 1918 (Yergin 1991).[7]

Another source of conflict early in this period was the growing competition between the Majors for market share. Thus a series of price wars[8] broke out in various regional markets (Penrose 1968) culminating in a major conflict between Shell and Exxon in India during 1927 and 1928. Such price wars were the inevitable outcome of an industry characterized by an oligopolistic structure and a tendency to carry excess capacity because of the discovery of relatively large fields whose production attracted significant economic rent to produce oil.[9] Furthermore it was (and still is) an industry with a cost structure characterized by very low short-run marginal cost which meant marginal revenue always exceeded marginal costs encouraging greater production (Stevens 2008b). In short it was an industry, if left to itself, which would suffer from chronic overproduction leading to intense price competition to ensure market outlets.

However, all of this conflict between the Majors was taking a serious toll on their profitability and effectively encouraged them to seek a collaborative solution. The main sources of collaboration first emerged in 1928. In that year, the heads of Exxon, BP and Shell met at Achnacarry Castle in Scotland where they effectively carved up the international oil market based upon the 'As-Is Agreement'. This agreement, which quickly included the other Majors, remained secret until 1952. In essence it agreed, as the name implied, to maintain existing market shares and, above all, not to compete on the basis of price. To this end the Gulf Basing Point System was introduced (Penrose 1968). Thus irrespective of geographic location, oil products were priced on a cost insurance freight (c.i.f) basis as though they had originated on the Gulf of Mexico that is, based on the US domestic price and the landed price equalized by means of a 'phantom freight rate'. In 1944, following complaints by their respective navies to the US and British governments, a second basing point was introduced at the Abadan Refinery in the Persian Gulf. This effectively opened up world oil markets to low-cost Middle East oil without threatening the price structure. This was one of the most important decisions for oil in the twentieth century since it allowed cheap oil to fuel the post-Second World War economic boom. Another example of the growing collaboration between the Majors in this period was the Red Line Agreement, also signed in1928. This was an agreement between the co-owners of the IPC (that effectively included five of the eight Majors) to not compete for exploration acreage in the region roughly coinciding with the old Ottoman Empire.

A post-imperial world: 1945–70

The relationship between the major oil-consumer governments – the states in the Organisation for Economic Cooperation and Development (OECD) – and the oil companies in this period was one of collaboration. The Majors were able to supply the ever-increasing demand for oil, which was fuelling the 'OECD economic miracle', and to do so with falling real prices. The result was that the US and British governments were content to leave oil issues to the Majors and the civil service was specifically instructed to let the Majors get on with the business of fuelling post-war reconstruction and eventual boom in the 1960s. As will be seen in the next period, this changed dramatically with the oil shocks of the 1970s but one of the reasons the 'shocks' were shocks was because having left oil to the Majors, no one in government – politicians or bureaucrats – had the first idea over what was going on in the industry in the 1970s as international prices quadrupled in the first oil shock of 1973–4 and quadrupled again in the second oil shock of 1978–81.

In terms of the relationship between the Majors and producer governments, the period was characterized by conflict. The key source of this conflict was the division of the rent from oil. This is based upon the agreement between the owner of the resource (normally the state represented by the government) and the operator (frequently a multinational corporation). The agreement defines property rights, roles and responsibilities. As outlined at the start of the chapter such agreements emerged and operated post-Second World War in the context of three interconnected cycles: the political cycle, the resource nationalist cycle and the obsolescing bargain cycle. As will be explained, it is the interactions of these cycles that determined the extent to which the various relations were closer to conflict or collaboration.

The political cycle had two dimensions – international relations and economic. The international relations dimension concerns the rise and pursuit of the Cold War between the US and the Soviet Union. Thus the world became divided into spheres of influence based upon client–patron relationships. As will be seen, this division began to seriously affect relations between some producer governments (individually and collectively within OPEC) as representatives of the 'Third World' and the Majors as 'representatives' of Western capitalism.

A major driver of this cold war conflict was the rapidly increasing demand for oil. The 1960s saw the 'OECD economic miracle'. The US, Western Europe and Japan grew at an unprecedented rate. With an income elasticity of demand for oil of one, this meant very strong demand growth for oil met largely by imports. Between 1958 and 1972 world oil demand (excluding communist areas) grew from 16.5 to 46.3 million b/d, an annual average growth rate of 6.3 per cent while between 1965 and 1970, the annual average was 8.1 per cent growth. In the same period, non-OPEC production grew at only 6.8 per cent thus greatly increasing OPEC's share of the market and its market power. By 1973, OPEC produced 53 per cent of world production outside of the Soviet Union. This raised the strategic importance of the Middle East to the West. Thus the region became one of the main battlefields of the Cold War with the US and the Soviet Union carving up the region to secure their own client states and trying to subvert the client states of the other. As will be seen below, it also inevitably increased the bargaining power of producer governments in the Middle East.

The economic dimension of the political cycle refers to the view – usually driven by a mixture of ideology and economic crises – as to the role of government in the allocation of economic resources within an economy. Thus if it is believed that markets do not work efficiently because of market failure, governments should intervene. This period was one in which, outside of the US, the generally accepted view was for government intervention although the degree of such intervention was debated (Stevens 2008b). However, as has already been explained, the relationship between OECD-consuming governments and Majors was based upon a view of the political cycle which suggested the Majors were doing 'a good job' and the industry should be left to work things out in the context of markets without undue government interference.

The story of relations between producer government and companies was very different. Here the other two cycles – resource nationalism and the obsolescing bargain become crucial. As defined in the introduction, the resource nationalist cycle refers to the desire of the government to assert control over its natural resources together with its attitudes to foreign companies. Thus the peak of the cycle would be a strong desire for control with xenophobic dimensions. This cycle tended to be driven by the magnitude of and need for oil revenues, which to a large extent was determined by oil prices. The obsolescing bargain cycle refers to the idea that the terms of the agreement are determined initially by the relative bargaining power of the parties at the time of negotiation. However, after the operator has made the necessary investments, the bargaining power swings strongly to the government leading them to seek renegotiation of terms. This cycle is driven in part by the resource nationalist cycle, the size of the eventual oil discovery and the price of oil.

This period, the post-imperial world, was characterized by several elements. First, it was obviously a postcolonial world where all forms of nationalism were in the ascendancy as newly independent nations tried to assert their sovereignty. This was in a context where state intervention in the economy was increasingly regarded as the norm. This was especially true for oil, which was seen as a key 'commanding heights' sector by the producing governments. In this period, the influence of the Soviet Union and its command economy approach to resource allocation was especially powerful in developing countries who were seeking to break out of their circle of poverty.

Second, there was the rise of the concept of 'permanent sovereignty over natural resources'. The United Nations passed its first resolution on this issue in 1952. In 1962, a resolution recognized the rights of a country to dispose of its natural wealth in accordance with its national interests. In 1966, resolution 2158 was even more explicit and host countries were advised to secure maximum exploitation of natural resources by the accelerated acquisition of full control over production operations, management and marketing. This effectively legitimized the concept of the obsolescing bargain.

Third, there was growing dissatisfaction with the oil concessions signed in the previous period, especially in the Middle East. Four issues dominated. First was the very long life of the original concessions. In Iran, Iraq, Kuwait and Saudi Arabia the average life was 82 years. Second, the areas covered by the agreements were huge. In the four main countries, 88 per cent of the national area was covered including all of Iraq and Kuwait.

Furthermore there were no relinquishment clauses. The Majors could simply sit on acreage, including commercial discoveries, and do absolutely nothing. Third, there was growing dissatisfaction with the fiscal terms. This first surfaced in Venezuela in 1948 when the government insisted (successfully) on a profits tax in addition to the lump sum royalty. This idea rapidly spread to the Middle East and by 1952 all the major countries in the region had switched to a system of profit taxes.[10] This new fiscal system created a series of major disputes over the setting of the posted prices used to compute revenue and over what should be included as costs and how they should be treated.[11]

The final and main source of dispute was that the concession gave the Majors total managerial freedom within their concession areas. Given the size of these areas, this gave them the power of being a 'state within a state'. They could choose the rate of exploration and development plus production levels. The producer government could try and influence these decisions but their power was strictly limited. A good example was how the Iraq Petroleum Company (IPC) treated the government of Iraq. The Majors had to manage oil supplies from various sources to balance the international market and protect prices. Because the IPC had the widest ownership outside of Iran (five out of the eight Majors), it became the swing producer (Blair 1976). If markets were oversupplied, IPC would pick a fight with the Iraqi government who then 'pressured' the IPC to reduce output. If the market was under-supplied IPC offered 'concessions' and production increased. Furthermore, there was nothing in the original agreements to allow governments any role in the operating companies.[12]

These drivers all fed the growing 'resource nationalism' which expressed itself most obviously in popular opposition in the 'Arab Street' to the role of the Majors (Hirst 1966). With this context the battles between producer governments and the Majors commenced. The first salvo was the Iranian nationalization of 1951, which was undermined over a period of about two years by the boycott imposed by Anglo-Iranian and the eventual intervention of the Western powers leading to the overthrow of Prime Minister Mossadegh.[13] These events dampened enthusiasm among other producer governments for the pursuit of the nationalization route to secure greater control.[14]

The next battlefield was the producer governments seeking to secure greater oil revenues. This led to a two-pronged attack. First, there was pressuring the Majors to produce more oil and second to renegotiate the fiscal terms. Furthermore, this was in a world of falling real oil prices. Initially, the Majors who were administering the posted prices tried to maintain the traditional link between US domestic oil prices and these prices. This proved impossible as low-cost Middle East oil began to gain market share. Unilateral cuts by the Majors to posted prices that reduced producer government revenues led in 1960 to the creation of OPEC. Thereafter, although posted prices of existing crude were effectively frozen, new crudes were given relatively low-posted prices and the actual realized prices of arms length deals was falling (Adelman 1972).

The governments had success over improving fiscal terms not least by getting agreement that royalties should be expensed (Seymour 1980). Such improvements were boosted by the introduction of new type of upstream contracts such as joint ventures, production-sharing agreements and service contracts (see Chapter 8). These allowed new international oil companies to enter the fray, but as such contracts spread, even the Majors began to accept these terms as the norm. A key characteristic of these new contracts was the progressivity of the fiscal terms that greatly strengthened the ability of the producer governments to capture rent. Also, the rising demand for oil helped absorb some of the pressure for increased production although there was intense competition between producer governments trying to secure greater output.

However, despite these gains by the producer governments, the underlying problem of managerial freedom for the IOCs and lack of producer-government control over operations failed to be addressed. The result was a growing upsurge of 'resource nationalism' aimed at governments taking control of the Majors' operations. To this end, there began the creation of national oil companies (NOCs), which became a central part of the 'resource nationalism' story.[15] As the tensions and conflict grew this translated into growing pressures for the producer governments to nationalize the Majors. While the fate of Mossadegh discouraged the rulers from pursuing such a path, by the time of the June War of 1967 the pressure reached a peak. To try and defuse the situation, Zaki Yamani, the oil minister of Saudi Arabia proposed the concept of 'participation' to begin the process of seizing control without provoking a political backlash from the host governments of the Majors. This led to negotiations which led to the October 1972 General Agreement on Participation giving producers an initial 25 per cent equity stake, projected to rise to 51 per cent by 1982 (Stevens 1975). Meanwhile, these pressures from the producer governments on the companies shifted the international oil markets into the next period of the 'oil shocks'.

The second set of relationships that needs to be considered in this period is that between the companies themselves. For the relationship between the Majors, the period represented a peculiar mixture of competition and collaboration. The cosy cartel-type relationship that emerged for the Majors from the 1928 'As-Is Agreement' came under ever-greater scrutiny especially from the US government. One solution was for them to convert their operating companies – the Iranian Consortium, the IPC, Kuwait Oil Compnay (KOC), Aramco and so on – into joint ventures. In this way the constraints of anti-trust legislation preventing the sharing of information was overcome by virtue of the fact that each Major knew what the others were doing and planning to do since the information was being shared in the relevant operating companies' board rooms. This collaboration enabled the Majors to effectively control the international oil industry for nearly 30 years. The result as already indicated was plentiful supply at declining real prices.[16]

However, the Majors began to face competition from two other sources. The first was the Soviet Union, which began to export crude and products into western Europe where they were able to undercut the Majors' control of the price.[17] The second was the new companies – the so-called Independents – that were a mixture of small American independents (Phillips, Amoco, Occidental etc.) seeking access to low-cost Eastern Hemisphere crude and the national oil companies of some European countries (ENI, Repsol etc.). The significance of this was that after the US limited crude imports into the US in 1959, these companies were forced to compete in European product markets to find an outlet for their stranded crude. They did so by cutting prices. This began seriously to damage the real profitability of European refinery operations.[18]

The result was the Majors saw these independents as a serious threat. As will be seen in the next period, the opportunity to damage the Independents arose in Libya. Colonel Ghaddafi's demands for price negotiations threatened the operations of the Independents since Libya was their only source of eastern hemisphere crude. Initially the companies collaborated by creating (with the US government suspending the Anti-Trust implications) the Libyan Producers Agreement to prevent Libya from picking off the weaker companies. However, the Majors quickly realized that higher Libyan prices would damage the Independents and so they acceded to Libyan demands for higher prices not realizing the extent to which they were opening the floodgates for the next period.

The oil shocks: 1970–86

The result of the battles outlined in the previous period had led to an unprecedented victory for the producing governments. The resource nationalist and obsolescing bargain cycles were moving towards their peak. In 1970, on the fiscal front, as described previously, following huge pressure from Libya, the Majors accepted producer-government involvement in setting posted prices creating a bilateral negotiating process. By October 1973, the governments took over that prerogative and unilaterally announced increases to price in the context of the Arab Oil Embargo that brought politics into the perceptions of the international oil market on a grand scale. This was the first oil shock. In 1978, the Iranian oil workers went on strike, which was a significant factor in the subsequent overthrow of the Shah and the Iranian Revolution. Together with the subsequent Iraq-Iran war this led to the second oil shock and yet another quadrupling of crude prices. With the first and second oil shocks, oil revenues for the producer governments reached unprecedented levels. Apart from boosting the resource nationalist cycle by making the Majors seem unnecessary to the governments, this also greatly increased the role of the producer governments in their own economies reinforcing the political cycle of government intervention in resource allocation.

On the issue of control and managerial freedom, in 1972, Algeria and Iraq opted for the nationalization route. The lack of a response from the Western powers meant the Participation Agreement of 1972 was destroyed in a series of re-negotiations led by the aggressive stance of the Kuwait National Assembly.[19] By 1976, the old style concession in the Middle East especially had been swept away; the producer governments had full control over their oil operations and indeed their oil prices.

One consequence of this switch in control was that the producer governments as disposers of the crude now had the choice of where to locate refineries. During the 1950s and 1960s refineries had increasingly been located in the major markets of the OECD.[20] The newly empowered NOCs of the OPEC countries began to consider developing their downstream capabilities by building export refineries. However, the dramatic increase in spare refining capacity following the fall in oil demand after the two oil shocks made such a move commercially unattractive.[21] Eventually however, this spare capacity began to be offered for sale leading to a number of purchases by the NOCs, which in turn, in some cases, led to joint ventures between the NOCs and the IOCs.

In this period, a new source of tension emerged with the creation of the International Energy Agency (IEA) in 1974. This was the brainchild of Henry Kissinger who was perceived (in the Arab world at least) to be extremely hostile. It had been created as an explicit counterweight to OPEC and its first acts were to create an emergency-sharing scheme with a view to protecting OECD oil importers from another Arab oil embargo. OPEC regarded the IEA with deep suspicion, especially the Arab members who viewed it as an explicit means for the consumers to reverse the revenue gains they had made following the two oil shocks. This view of potential conflict was reinforced following meetings of the (then) G7 in Venice and Tokyo that clearly indicated that oil imports would be targeted, and especially those from OPEC.

Following the second oil shock of 1978–81, yet another new source of conflict emerged, this time between the producer governments that made up OPEC. The problem was triggered because of the problem of excess capacity to produce crude which had dominated the industry since its inception.[22] This excess existed for three reasons. First there was significant rent in the oil price that created a strong incentive for the owner of any oil-in-place discovered to convert it into producing capacity. Second the industry was prone to 'accidents' which took out of play capacity, which required the rapid development of replacement capacity. Once the 'accident' had been resolved the extra capacity became surplus to requirements. Finally, the industry was and is an industry of consensus. Everybody believes the same thing at the same time; observes the same signals and behaves accordingly. For example, at the end of the 1970s the consensus was that oil prices would rise steadily forever, encouraging IOCs to invest in exploration and development. This led to the rapid rise in non-OPEC production that forced OPEC to close in capacity. By 1985, over 40 per cent of OPEC's capacity was closed in (Stevens 2008b).

The history of the oil industry has been about managing the spare capacity. As already outlined, between 1928 and the late 1940s control was achieved by the 'as-is' agreement and in the 1950s and 1960s by the joint-venture structure of the main operating companies. However, after the nationalizations between 1972 and 1976 it was OPEC that was forced to play the role of market controller and contain excess capacity. Between 1974 and 1978, effectively Saudi Arabia played the key role by acting as swing producer to balance the market. This was helped by the fact that a number of OPEC countries decided in this period to restrict capacity expansion on the basis that 'oil under ground was worth more than money in the bank', that is, in expectation of rising future prices.[23]

However, following the second oil shock of 1978–81, the subsequent collapse in oil demand and the rise of non-OPEC supply,[24] OPEC's control problem became acute as their spare capacity rose. In March 1982, for the first time OPEC tried to introduce pro-rationing in much the same way as the US had done before the Second World War. However, this required the members agreeing on how much OPEC oil markets required and then allocate this 'call on OPEC' by means of quotas. Quota determination proved to be (and indeed still is) a major source of conflict both in terms of the allocation and in terms of the extent to which quotas were respected or ignored.

In general, a combination of error[25] and extensive cheating by some OPEC members threatened this control mechanism. However, for a while, Saudi Arabia absorbed this by reducing its own production, thereby trying to defend the price of $34 per barrel agreed within OPEC in October 1981 as the marker price. In 1980, Saudi Arabia produced 10.3 million b/d but by 1985 this had fallen to 3.6 million b/d.[26] However, this caused considerable problems within the kingdom. The senior princes were dismayed at how the lower oil revenue had caused serious fiscal problems for the kingdom and their own personal finances. At the same time the oil techno-structure felt the policy to defend higher prices was misplaced since it was causing oil to be pushed out of the global energy mix, thereby threatening the future value of the country's very considerable reserves.[27]

In the summer of 1985, the result was a fundamental shift in Saudi policy to move to lower stable prices. To this end, the policy was to maintain a cushion of spare capacity (at least 2 million b/d) to mitigate any outage leading (again) to a price spike. The result was the 1986 oil price collapse as Saudi Arabia refused to compensate overproduction from other sources and announced a policy of moving from administered prices to netbacks. This third oil shock created a whole new set of circumstances in the industry in the context of conflict and collaboration that were to have very significant impacts in the following period.

Markets appear to work: 1986–99

The oil price collapse of 1986 marked a major change in the resource nationalist and the obsolescing bargain cycles. Lower oil revenues prompted some governments to turn again to the IOCs in the hope that their involvement would offset the lower oil prices with higher volumes. The result was increasing supplies, which continued the struggle within OPEC over quota discipline. Arguably this was a major explanation behind the Iraqi invasion of Kuwait in 1990.[28] Certainly Kuwait had been foremost among those overproducing after 1985.[29] The view was that this was an explicit policy by Kuwait to undermine oil prices to weaken both Iraq and Iran in the aftermath of the Iraq-Iran War that had ended in August 1988.

The aftermath of the 1990 invasion and the subsequent liberation was to carry important implications for the oil markets of the world. It sowed the seeds for the 2003 invasion, which, as will be explained below, was a major factor in terms of conflict and cooperation in oil markets in the next period. It also introduced a degree of complacency over oil markets within the oil-importing countries. For some time there had existed a 'doomsday scenario' within the corridors of the IEA.[30] This was some form of military conflict that would threaten the supplies of oil coming out through the Straits of Hormuz in the Persian Gulf. This concern began to emerge with the outbreak of war between Iraq and Iran in 1980 and was compounded as both sides began to conduct a 'tanker war' with Kuwait tankers feeling particularly vulnerable to Iranian attacks. However, a degree of intervention by the Soviet Union and the US contained the problems and while it was claimed by Lloyds of London that 546 commercial ships were attacked, oil supplies remained unaffected. However the real test came with the invasion of August 1990 and the subsequent military action. In the event, although oil supplies were constrained as the result of an oil embargo imposed by the United Nations, prices only responded to a very limited extent. The view that was peddled, especially by the OECD countries and the IEA, was that this was because markets had been allowed to work. Such views fitted perfectly the 'Washington consensus', which by now was at the height of its popularity. However, there was a clear failure of such views to realize that a major reason for the limited price response was the already-mentioned Saudi willingness and ability to carry excess capacity to produce crude and use it to stabilize oil markets. This failure of understanding became important as the neo-conservatives' agenda in the US began to consider the need to dispense with Saudi Arabia and promote a new landscape in the Middle East that led ultimately to the 2003 invasion of Iraq and the subsequent shambles for the whole region that was to become its legacy.

Meanwhile a new phenomenon began to emerge in this period and that was the rise of paper trading markets for crude and products. In October 1978, there was only one small futures market for gasoil in Chicago. Otherwise none of the paper market developments that had characterized other commodities were present.[31] The reason for the failure of such markets to develop was that their main justification – to hedge price risk – was not an issue in international oil markets where prices were administered, initially by the Majors and after 1973 by OPEC. The paper markets for oil first emerged in the early 1980s in the form of an informal forward market. However in 1987, the New York Metal Exchange (NYMEX) opened the first official regulated trading floor based upon West Texas Intermediate (WTI). In 1988, the International Petroleum Exchange in London followed suit by trading in Brent. One consequence was that these paper markets began to aggravate price volatility, although this view is controversial. Whatever the reality, price volatility did become a source of conflict in the next period over the rise in prices experienced after 2002. The OECD oil importers blamed OPEC for inadequate supply leading to higher and more volatile prices while OPEC blamed the 'speculators' in the paper markets.[32] As will be seen below, in the later period, the causes of higher prices have proved to be a major dispute not only between producers and consumers but also between analysts.

This period also saw the rise of growing competition between the companies. A major problem facing the IOCs was that their financial strategy was based upon a system of 'value based management'.[33] Put simply, this argued that if a company could not earn a rate of return on its capital employed at least as good as the rest of the sector, it should give the funds back to the shareholders. Throughout the 1990s this presented a major challenge. In particular the IOCs faced the problem that the bulk of the low-cost reserves in the world were off-limits. At the end of 1998, 53 per cent of world proven reserves were in Saudi Arabia, Iraq, Iran and Kuwait. The first two were off-limits politically and while the latter two were trying to allow IOC entry, in both cases the process fell foul of domestic politics and, in the case of Iran, was not helped by US sanctions as a result of the Iran Libyan Sanctions Act passed in 1996. This obsession with maximizing shareholder value also put the IOCs at a major disadvantage in securing upstream access vis-a-vis the increasingly aggressive NOCs from the major oil-importing emerging market economies such as India and China given their (relatively) limited interest in the costs of capital. In this period, these NOCs began to look for access to equity oil abroad giving rise to talk of a new 'scramble for Africa'.

From these developments a number of consequences for conflict and cooperation began to emerge. First, they led in the late 1990s to a series of significant consolidations as there was merger frenzy between the IOCs kicked off by BP's acquisition of Amoco in 1998. This merger frenzy was also encouraged by the oil price collapse of 1998 (see below) that meant the companies could be bought at relatively low costs and that the price/cost per barrel of reserves was cheaper than actually trying to find it at the end of a drill bit. Second, it meant that the level of investment by the IOCs in all stages of the industry flagged, which sowed the seeds for the serious capacity constraints that dominated the industry in the next period. In particular there developed a serious shortage of refinery upgrading capacity which meant that sweet light crude increasingly traded at an ever-widening differential to heavy source crude.

Meanwhile, conflict within OPEC also continued to characterize the period after the liberation of Kuwait in 1991. Disputes over cheating on quotas continued to rumble on. In particular, Venezuela through its policy of 'apertura' to encourage IOC entry was guilty of completely ignoring quotas and producing as much as possible. Between 1988 and 1998 Venezuelan production increased from 2 to 3.5 million b/d. Iran was also blatantly ignoring quotas. At the OPEC meeting in Jakarta in October 1997, OPEC hopelessly overestimated the call on its oil as a result of a failure to account for the Asian financial crisis, the very high levels of overproduction by OPEC and the return of Iraqi crude under a UN humanitarian programme. The result was the 1998 price collapse.

As the price began to weaken in the second quarter of 1998 there were attempts by some non-OPEC countries such as Mexico and Russia to give support to OPEC but the oversupply was too large. By January 1999, the price of Brent fell below $10 per barrel. The price was rescued by two unconnected political events. In Saudi Arabia, Crown Prince Abdullah was given de facto control as King Fahd effectively moved into retirement forced by ill health. Top of Abdullah's policy agenda was détente with Iran for fear that the US might prove to be an unreliable ally. Thus the stand-off between Iran and Saudi Arabia over Iran's failure to stick to quota, which was a major factor behind the collapsing price, was solved by Saudi acquiescence. In Venezuela Hugo Chavez became president and it was clear he would reverse the 'apertura' policy and bring the country back into the OPEC fold.

Agreement was reached in March 1999. OPEC stuck religiously to the newly agreed quotas and prices began to recover. The benefits within OPEC of cooperation rather than conflict became clear. Subsequently, the quota issue disappeared as the call on OPEC crude rose as a result of dramatically growing demand, especially in Asia, and the level of spare crude producing capacity fell close to zero in 2004. The quota issue did re-emerge in the aftermath of the near collapse of the financial system following the bankruptcy of Lehman Brothers when oil demand growth became negative in 2008 and 2009.[34] Thus OPEC was forced to cut, to try and absorb the very large stock overhang that, as in 1998, dominated the market and threatened prices.

The cycles begin to swing again: Post 2000

As the industry began to move into the twenty-first century, the various driving cycles began to change. Following the collapse of the Russian economy in the summer of 1998, elements within the World Bank began seriously to question whether the free market philosophy that characterized the 'Washington consensus' was actually valid. After all, this was a country that had followed all the prescriptions of the 'consensus' to the letter and yet still had collapsed economically. Other problems in the energy field also gave rise to concerns over whether market forces were the ideal solutions. Perhaps, most spectacular was the serious electricity supply problems in California in 2000–1.[35] The result was a general mood in oil that the laissez faire attitudes of the 1990s were no longer appropriate and governments should intervene to avoid product shortages and the growing levels of price volatility. This view of the need for intervention was given important boosts by the growing concern over climate change and security of supply issues, most obviously after 9/11 and in the aftermath of the 2003 invasion of Iraq.

At the same time, the resource nationalist cycle began to re-emerge. Several factors explain this phenomenon. The revival began in Latin America and Russia but for rather different reasons. In Latin America, successive generations of the 'wretched of the earth' had seen the international companies taking 'their' oil and minerals and they had seen little or no benefit. As a result they blamed the companies. This was rather unfair since the companies had been paying lots of taxes to the relevant government but these kleptocracies had been misappropriating the revenue in classic 'resource curse' fashion.[36] Thus when populist politicians promised to ensure the benefits from oil and minerals trickled down, they were elected. In Russia, resource nationalism re-emerged as a growing reaction to the disgraceful sell-off of Russian state oil interests in the early 1990s as part of the programme of perestroika. All this had achieved was to produce a generation of oil oligarchs and no benefit whatsoever to ordinary people. Again, such a context encourages populist politicians to promise redress.[37] The result was growing hostility to foreign companies. In a process of contagion, such views began to re-emerge in the Middle East and North Africa.

The rising level of oil prices also helped this resurgence of resource nationalism, as skyrocketing oil revenues made IOC assistance seem much less urgent. Thus the obsolescing bargain began to re-assert pressure on the IOCs. At the same time, many producer governments began to rethink their depletion policies and argued (in a variation on the mantra of the 1970s described earlier) that oil in the ground was worth more than money in the bank, especially if the bank was plagued by toxic debts. The consequent slowing down of investment in capacity by many NOCs helped to generate the tighter market which greatly contributed to the rising price of oil which reached a peak of $147 on NYMEX early in July 2008. These constraints on capacity in all stages of the industry were reinforced as the result of inadequate investment by the IOCs. Because the tightening of fiscal terms made it even harder for the IOCs to justify investment, huge sums of money leaked out of capital pot for the private industry being returned by the IOCs to their shareholders through higher dividends or share buybacks (Stevens 2008b). This was reinforced because the competition for upstream access with the Asian NOCs that began in the earlier period was intensifying. Many observers began to talk openly about a scramble for resources. Indeed one of Shell's scenarios launched in 2009 was given the title 'scramble'.

Meanwhile, another source of conflict over oil began to move up the agenda. As prices began their seemingly inexorable rise to $147, as already outlined, there was serious disagreement between the main OECD-consuming countries and OPEC over the causes both of higher prices and their volatility.[38] The result was that in June 2008 a meeting was convened in Jeddah between the consumers and the producers in an attempt to agree on the causes of rising prices and their volatility and to seek solutions. A similar meeting in London in December 2008 followed. The basic disagreement was never solved although following the June meeting in Jeddah, Saudi Arabia did pay lip service to the supply issue by offering to increase production to mute higher prices.

The subsequent collapse in oil prices – by December 2008 the OPEC basket reached $38.60 – did rather mute the resource nationalist cycle. Many producer governments were now facing severe financial problems. Some tried to turn back to the IOCs to seek greater production volumes but the recent history of the obsolescing bargain muted any interest from the IOCs.[39] In any case, the global economic recession meant spare capacity to produce crude was rising rapidly which yet again resurrected conflict over OPEC quota discipline.

In this period, undoubtedly one of the most controversial issues in the range of conflict versus cooperation concerned the invasion of Iraq in 2003. The controversy revolved around the motives behind the US-led invasion. Many simply assume that the invasion was caused by a US desire to secure Iraqi oil for its own use and to replace Iraq as the oil market controller thereby undermining the position of Saudi Arabia. Such a view misses a number of important points. First, there were a great many other reasons for the invasion, none mutually exclusive. These range from the neo-conservatives' agenda to transform the Middle East into a sea of Western market capitalism underpinned by democracy; a genuine belief among many in the various intelligence communities in the likely (or forthcoming) existence of weapons of mass destruction; concerns that a US military presence was undermining the legitimacy of the Al Saud but removal of such a presence was unthinkable while Saddam remained in power; a long-standing feud between the Bush family and the Al Takrit; and the list could go on and on. Second, if the US really did want Iraqi oil, there were a lot cheaper ways of securing it than going to war, something which the developments in Libya illustrated. Third, while it is true that oil interests dominated the presidency of George W. Bush, these 'interests' were in fact those of the US domestic industry. The last thing wanted by them was cheap oil flooding the market bringing down the price (domestic and international) of crude oil.

Following the oil price collapse in the second half of 2008, OPEC struggled to tighten the market. In particular it was trying to absorb the growing stock overhang, which emerged as oil demand fell in 2009 as a result of global economic recession. OPEC discipline was good and the market did begin to tighten and prices began to recover. At the end of 2010, markets had become relatively tight and the expectation was that prices would move higher. The year 2011 saw the Arab Uprisings sweep the region (Stevens 2012a). These events removed small amounts from the physical market with the loss of Yemen, Libya and then Syria further tightening the markets. At the same time, the events scared the paper markets and prices continued to rise as there were fears (almost certainly unfounded) that the major oil exporters of the Gulf Cooperation Council (GCC) countries would follow the path of Tunisia, Egypt and Libya in some sort of contagion effect. However, a revival of fears of a renewed global recession following a succession of crises in the eurozone placed a lid on prices. Thus prices bounced between what was happening in the Arab Street and what was happening in the global financial markets. Into this maelstrom the Iranian nuclear issues moved dramatically up the agenda. Sanctions on Iran were tightened. In particular, a very ill-thought out embargo on Iranian oil exports was announced by the European Union (Stevens 2012b). This led to a significant rise in price as a result of natural transitional friction as European importers of Iranian crude scrambled to find alternative suppliers and markets considered what the possible Iranian response might be.[40]

Thus the current situation in oil markets is one of extreme uncertainty. This is not least because of great uncertainty in energy markets generally. Thus the prospects for nuclear power are uncertain following the tragic events in Fukushima as well as the prospects for a possible dramatic increase in the supply of natural gas following the shale gas revolution in the US; questions are being raised about the cost of renewables and the whole climate change agenda and its influence on energy policies are highly uncertain.

Notes

1. Otto von Bismarck once famously remarked 'Only a fool learns from his mistakes. I learn from other peoples' mistakes!'.
2. This claim is disputed in Baku, Azerbaijan, which claimed F. N. Semyenov drilled the first well ten years before Drake in 1849.
3. Throughout this chapter the modern names of the major oil companies will be used.

4. This process had begun in 1914 when the British government took a controlling interest in (what became) BP as a result of switching the Royal Navy from coal to oil fired ships.

5. Arguably, France was less active in chasing concessions and gave up its mandate in the Kirkuk Wilayet (granted under the Sykes-Picot Agreement of 1916) to accept a place in the IPC Consortium.

6. These were the agreements signed in Iran in 1901 which created BP, in Iraq in 1928 which created the Iraq Petroleum Company (IPC), in Kuwait in 1933 which created the Kuwait Oil company (KOC) and in Saudi Arabia in 1933 which created eventually the Arabian American Oil Company (Aramco) (Stocking 1970).

7. It is perhaps significant that it was Germany that developed coal-to-liquids technology via the Fischer-Tropsch technology as an earlier effort to try and mitigate this lack of access.

8. A price war is different from price competition insofar as a price war normally has some sort of specific strategic objective (see Penrose 1968: 55).

9. There were other reasons why the industry has been characterized for much of the time by an excess capacity to produce crude oil. These will be enumerated below.

10. It is interesting to note that the US government actively sponsored this fiscal deal between Aramco and the Government of Saudi Arabia. This transferred revenue from Washington to Riyadh and was effectively a means to give support to the Al Saud.

11. For example, a major dispute was how royalties should be treated (Seymour 1980). The conflict over posted prices arose because the operational vertical integration of the Majors meant there was no market price for crude. To be able to compute revenue and hence profit for tax purposes meant the Majors unilaterally set a price which was supposed to reflect a market price. The unilateral nature of this decision was to prove a major source of conflict between the Majors and the governments until 1973. The first oil shock was effectively the producer governments wresting the pricing decision from the Majors.

12. Some did have clauses allowing the government to purchase shares in the event of any being sold. However, since no equity was ever offered this was totally ineffective (Stevens 1975). Also the original IPC agreement gave the government some involvement but this was 'sold' by the government of Nuri Al Said for a small sum following the overthrow of the monarchy. Giacomo Luciani in a private communication has suggested this illustrates three things first, that personalities matter; second that the political leadership of the producing countries could be aware of the importance of oil; third that the companies were extremely short-sighted in the defence of their interests.

13. It is now a matter of historical fact that the coup that overthrew Mossadegh was engineered by the CIA and Britain's MI6 (Bamberg 1994).

14. In the aftermath of the coup that overthrew Mossadegh, he was sentenced to death. While the sentence was never carried out this remained a powerful image for governments who might have contemplated a more aggressive route to control their oil.

15. The NOC story can only be briefly referred to in this paper due to constraints of space. There is however an extensive literature on NOCs which tends to be of two types – either fairly general or studies of specific NOCs. A good overview of the debates and issues can be found in Canadian Department of Energy (1972);

Madelin (1974); Krapels (1977); Noreng (1980); UNCRET (1980); Grayson (1981); Robinson (1993); Van den Linde (2000) and Marcel (2006). The specific NOC studies are widely different reflecting the obvious point that the (OPEC) NOCs 'are as diverse as the member countries themselves' (Hartshorn 1993:165). For differences in the legal status of NOCs see also Bentham and Smith 1986; Bentham 1988. A major new work on NOCs comes out of a major study undertaken by Stanford University (Victor et al. 2012).

16. In theory and practice the Majors could have pushed for much higher prices. The fact they did not was the result of two constraints. First they were concerned about competition to oil in the global primary energy mix, especially from Nuclear. Second they were concerned about any potential backlash from their host governments if oil began to become more expensive.

17. American oil Majors began to complain about the 'communist technique of price cutting'! (Penrose 1968).

18. European refineries were making accounting losses throughout this period but this was the result of transfer pricing by the Majors trying to minimize their global tax bills.

19. There is a widely believed view in some quarters that the US favoured these events because they wanted higher international oil prices. These would damage export competitiveness of their European and Japanese export competitors when the US balance of payments was a cause of concern. It would also give a boost to the flagging US domestic oil industry (Chevalier 1973).

20. Before 1945 the majority of refining capacity had been located on the oil fields reflecting the large losses incurred in relatively primitive refining plus the fact that the demand barrel was extremely unbalanced with Europe having plentiful coal looking for light transport fuels and Japan without coal looking to heavy fuel oil to raise steam under boilers.

21. This was compounded because the IOCs, expecting the demand strength seen in the late 1960s and early 1970s to continue, had commissioned a huge increase in refinery capacity. This capacity began to come on-stream just after the first oil shock started to process of demand reduction and destruction.

22. The excess capacity is defined as production which can be brought on-stream within days or at most couple of weeks.

23. It was also helped because several countries, notably Iraq and Libya, were reluctant to build up financial assets in the West for fear of expropriation.

24. The nationalizations left the IOCs with a need to find alternative sources of crude oil and the much higher price gave them the incentives and the financial means to do so.

25. To allow flexibility in the negotiations and allocation of quotas, OPEC often grossly overstated how much OPEC oil was required. This was in a context where data on demand and supply were (and still are) absolutely awful.

26. In August 1985, an average of six sources computed by the author put Saudi production at 2.3 million b/d.

27. Indeed in the early 1980s, Saudi Arabia had been deliberately setting prices below the rest of OPEC in an effort to restrain prices rises and their obviously damaging effect on oil demand.

28. There were of course other factors at work behind the invasion to do with internal Iraqi politics. However, another oil dimension to the invasion was the growing dispute between Iraq and Kuwait over the treatment of the Rumailah oilfield which straddled the border between the two countries.

29. In 1985 Kuwait produced 1.1 million b/d while in March of 1990 an average of six estimates computed by the author puts production at 2.0 million b/d.
30. In scenario building, a 'doomsday scenario' is one that has a very low probability of occurrence but, if it does happen, would be devastating to the players involved.
31. The first modern futures exchange emerged in 1710 in the form of the Dojima Rice Exchanges in Osaka.
32. They also blamed the higher prices on inadequate investment in refineries that put a premium on light sweet crudes, which were the headline prices on the paper markets.
33. This first started to emerge in the oil companies in the late 1980s and early 1990s. It was derived from financial theories such as the capital asset pricing model that were emerging from the universities and business schools.
34. It has also re-emerged as the result of Iraq signing a series of contracts with IOCs at the end of 2009 to boost its producing capacity. If the numbers are to be believed (which they should not) Iraq's capacity would be 12 million b/d within a few years. There is already talk within OPEC that Iraq should now come back into the quota which will lead to a major negotiation.
35. In reality the Californian crisis was not the result of markets. It was the result of a bungled deregulation when wholesale prices were deregulated and retail prices were not. The experience was described to the author as similar to a situation where a country that drives on the left hand side of the road decides to switch to driving on the right. And they intend to start on Monday with the trucks and the buses!
36. 'Resource curse' presents a very specific source of conflict within the context of the oil industry but at a national and local level. Unfortunately space limitations preclude a greater discussion of this issue (those interested might look to Stevens and Dietsche 2008).
37. Such politicians also had a hidden agenda which was to use energy resources to try and reassert the Russian political power lost with the collapse of the Soviet Union.
38. Formal relations between consuming and producing governments go back quite a long way. Following the first oil shock of 1973–4 the 'North-South Dialogue' was created although it failed to produce anything of note. Then in 1991 discussion was resumed under the auspices of the International Energy Forum (IEF) and in 2002 at the Eighth Conference an IEF Secretariat of was created to be located in Riyadh. In April 2001 the IEF launched the Joint Oil Data Initiative (JODI) whose function was to try and improve the quality of data for the industry (Stevens 2011).
39. An obvious exception to this was Iraq where there was huge interest from the IOCs. However, the initial offerings from Iraq were very unattractive and in the subsequent bidding round, terms for the IOCs had to be improved.
40. The prospect of the closure of the Straits of Hormuz again raised its head despite the fact that this was extremely unlikely (Stevens 2012a).

2
Minerals: Industry History and Fault Lines of Conflict

David Humphreys

This chapter examines the history of the minerals industry and draws out from this history some observations on sources of conflict in the industry. It also considers factors which have contributed towards periods of greater cooperation. Although the history of minerals displays some broad similarities with that of the oil sector discussed in Chapter 1, it also has some marked differences to which will be drawn attention in the course of the narrative. Importantly, however, a conclusion of the chapter is that the minerals sector shares with oil clear signs of a shift in recent years towards greater state interventionism and the adoption of state capitalist modes of organization.

For the purposes of this chapter, minerals are taken to be what are known in the business as 'hard minerals' to distinguish them from oil and gas. They include all products recovered by mining, including metals such as copper and gold, and non-metallic minerals such as diamonds and potash. They also include the mined energy minerals, coal and uranium. Clearly, this is a very wide range of products and there are limits to how far it is possible to generalize about such a diverse industry. To make the subject manageable, primary emphasis is given in the chapter to the major traded metals, although reference is made along the way to other mineral products.

The approach adopted is essentially chronological, being based around the broad historical regimes outlined in the introduction. The chapter starts with an examination of mining up to, and including, the Second World War, before progressing through the period of decolonization and the Cold War to the resurgence of economic liberalism in the 1980s and 1990s. It ends with a review of events since 2000 and some reflections on potential future sources of conflict in the industry.

Early industrialization and imperial liberalism

The use of metals and minerals goes back way before the industrial era, but it was industrial development which stimulated the large-scale demand for minerals and gave birth to the modern mining industry. Countries having local resources of minerals were naturally advantaged in the early stages of industrialization. Great Britain, the crucible of the industrial revolution, had local resources of iron ore, limestone and coal with which to build a basic iron and steel industry. During the first half of the nineteenth century, Great Britain was the world's largest producer of copper and lead, accounting for 40 per cent of global output of these metals, as well as 30 per cent of the world's tin. During the same period, Germany accounted for around two-thirds of world zinc production (Julihn 1928; Pehrson 1929; Smith 1929; Umhau 1932).

By the second half of the nineteenth century, the low quality of local resources of key minerals in Europe was restricting their supply and threatening to inhibit the process of industrialization. Gradually, the minerals industry started to go international, with mines opening up in southern Africa, in East Asia and in Latin America to serve the requirements of the growing industrial centres. In the 1850s, Great Britain was overtaken by Chile as the world's largest producer of copper and in the 1860s by the US. By 1880, Great Britain's share of world copper production had shrunk to under 2 per cent. Symptomatic of this shift in the location of mineral production, the London Metal Exchange was set up in 1877 for trading metals internationally, its standard three-month contract reflecting the shipping time to London for Chilean copper and Straits tin.

Mining goes global

Investment from the industrializing countries played an important part in the development of the new, richer resources, but it was not all a colonial affair. The Chilean copper mines were for the most part owned and run locally, although shipping was largely controlled by Europeans. Tin mining on the Malay Peninsula was almost entirely in the hands of local Chinese, and it remained so after the Malay States became a British protectorate in 1895. Rio Tinto's acquisition of a copper mine in southern Spain (from which the company took its name) in the 1870s could hardly be described as a colonial act. By contrast, tin mining in the Dutch East Indies (now Indonesia) was largely run by the colonial government while the mining of nickel in the French colony of New Caledonia, which commenced in the 1870s, was undertaken by French convict labour.

Colonialism and foreign investment played a much more prominent role in the opening up of Africa's resources. Capital raised in London and New York was important to the development of the diamond mines around Kimberley in the 1860s and 1870s and to the gold rush in the Transvaal that followed it in the 1880s. The push north from the Transvaal by Cecil Rhodes during these years into what was later to become Northern and Southern Rhodesia (now Zambia and Zimbabwe respectively) was driven by the belief that the area harboured substantial mineral resources, including diamonds and copper. Sharing a similar conviction, and intent on blocking the British push northwards, King Leopold II of Belgium gained control over the almost three million square kilometres of the Congo in 1883, an entitlement which was formally recognized by other European nations in the Act of Berlin in 1885. Things did not stop there, however, and in the next 15 years, in what became known as the Scramble for Africa, the entire continent of Africa, with the sole exception of Ethiopia, fell into the hands of one or other European power (Lynch 2002).

The colonization of Africa brought with it major mine development. Copper production started in Northern Rhodesia in 1906 although it was not until the late 1920s that the full extent of the Africa Copperbelt was appreciated and production really began to take off. The two principal companies involved were the Rhodesian Anglo American Corporation and Roan Selection Trust. To the north, the Belgian company Union Minière du Haut-Katanga (UMHK) embarked on the development of the copper deposits of the Katanga province of Congo, with first production taking place in 1911. In 1917, Anglo American was formed by Sir Ernest Oppenheimer and the US financier J. P. Morgan for the development of mines in southern Africa, the company taking its name from the regions where the company raised its initial capital.

In contrast to the situation in Europe, the US, with its large land mass and rich domestic resources, had less need to look overseas for mineral raw materials to support its industrialization. The pursuit of natural resources – and particularly gold – was, however, a key factor driving the settlement of the American West and was employed as justification for the appropriation of the lands of the Native American peoples as the new nation was pulled together and industrialized during the course of the nineteenth century.

Americans were, nevertheless, energetic businessmen and interested in financing mining projects wherever in the world there was money to be made, as could be seen from their involvement in South African mining. The mineral resources of Latin America had also caught their eye. Following the discovery at the Bingham Canyon mine in the state of Utah that copper could be profitably recovered from low grade ores if mining was on a sufficiently large scale, US investors, led by Daniel Guggenheim, acquired the El Teniente and Chuquicamata mines in Chile, in 1915 incorporating these interests into the Chile Copper Company. These mines were a few years later sold on to two of America's largest mining companies, Kennecott and Anaconda (Lynch 2002).

The growth of mining in Australia was, as with the US, driven significantly by local enterprise, this despite Australia remaining a British colony (or, more strictly, a collection of colonies) through to the beginning of the twentieth century. The industry was effectively launched with a gold rush in the colony of Victoria in the 1850s. However, while gold mining tended to be relatively small scale and local, it was the production of base metals that provided the launch pad for what were to become some of Australia's largest mining companies. The silver-lead-zinc deposits of Broken Hill in New South Wales spawned, in the mid-1880s, Broken Hill Proprietary (BHP) and North Broken Hill, and, in 1905, Consolidated Zinc. Further north, in Queensland, the silver-lead-zinc deposits of Mount Isa provided the foundation for the creation and growth of Mount Isa Mines (MIM).

While it is evident that major strides were taken in these years towards the development of the world's mineral resources and the establishment and organization of global markets in minerals, this cannot really be characterized as an era of collaboration. There were, no doubt, producers content to supply the metropolitan powers voluntarily in return for the opportunity to make profits, but there were other cases where resources were simply commandeered without reference to traditional landowners and where mine developments took place in enclaves in which labour had few rights and from which little benefit passed into the local economy. The colonial powers largely dictated the terms on which resources were developed and traded, backing this up with latent or actual force, so as to ensure that the needed commodities were able to flow back to the home nations and that the assets of investors were protected. Interference in this process from local political groupings (as with the Boers' resistance to the activities of the British in South Africa at the end of the nineteenth century) or from indigenous people (as with the aborigines in Australia or the tribal peoples of Africa) were not tolerated. The major competition was between the great European powers, and mining countries outside these regions were for the most part simply adjuncts of the big power game. This game came to a head with the First World War before breaking itself altogether in the Second World War.

Decolonization and the Cold War

The Second World War marked the effective end of the colonial era and in the years that followed most of the remaining European colonies were granted political independence. The focus of the global mining industry meanwhile was gradually shifting towards the emerging superpowers of the US and the Soviet Union, and towards developing countries. However, the economic relationships which had characterized the colonial era were not to be so quickly eradicated and many of the fault lines of conflict which developed during the colonial era persisted and indeed widened.

The early post-war years were a period of US geopolitical hegemony. The accompanying pressures for freer trade and cross-border investment meant that through the 1950s and 1960s big international companies – TNCs, or transnational companies, as they came to be known – from the US and Europe continued to play an important part in shaping international economic relationships, including those bearing on the minerals sector.

From the perspective of many newly independent countries, it seemed that these companies were simply perpetuating colonial relationships by other means. They served to keep key economic sectors in the hands of foreigners, they ensured that the rent from these sectors (as well often as the product) went back to the companies' home countries, and they generally sustained colonial-style relationships of economic dependency. Mining TNCs were commonly viewed as having little interest in the development of the economies in which they were operating (e.g. by processing their products locally) and many were suspected of operating schemes such as transfer pricing to avoid taxation in the host country. These relationships were, many countries determined, if not technically colonial, then decidedly 'neo-colonial'. This insight was reinforced by the fact that many of the mining companies focused their activities on areas of previous colonial influence. Thus French mining companies were deeply involved in North and West Africa (e.g. Algeria, Morocco, Mauritania, Niger, Gabon, Ivory Coast, Guinea), while British companies were prominent in mining investment in Australia, Canada and southern Africa. US mining companies, for their part, were making a big play in South and Central America.

With demand for minerals growing strongly as Europe reconstructed and Japan re-industrialized, the notion gradually began to take hold in many of the former colonies that, to break the cycle of impoverishment and dependency, political emancipation was not enough. To control their economic destinies and to capture the full value of their resources for their people, they needed outright ownership of their resource industries. Nationalized mining companies, it was reasoned, could be used to deliver economic and social benefits more directly to a country's citizens as well as serving as instruments of broader economic development. Not surprisingly, the notion had strong popular appeal. And it has close parallels with what was taking place at the same time in the oil sector. Nationalization not only promised the retention of wealth generated from minerals within the country but it constituted a clear and public rebuff to erstwhile colonizers and provided a cause around which politicians could rally their populations and assert their nationhood. In short, the fault line of this particular phase of conflict was emphatically and unmistakeably national. It was about the assertion of the sovereign right of an independent nation to organize and control its economic and political affairs. And it was about the use of national mineral resources for national economic development rather than for private, and foreign, profit.

The notion of nationalization as a means to attain greater public benefit from mining was not new and state ownership was already widespread in the industry. The Russian government had nationalized the country's mines, along with the rest of its industry, following the revolution of 1917, and all countries falling into the Soviet sphere nationalized their mining industries subsequent to the Second World War. Most of Western Europe's coal industry was also taken into state ownership. The Finnish government nationalized copper miner Outokumpu Oyj in 1924 while the Swedish iron ore producer, LKAB, was taken fully into state ownership in 1957. Among developing countries, the Indonesian government had acquired the country's principal tin mining companies on attaining independence from the Netherlands in 1950. Brazil created state-miner, CVRD, in 1942 and Bolivia nationalized its tin mining industry in 1952. The postcolonial period nonetheless saw a strong resurgence of interest in the idea of state ownership in mining.

Following the country's independence in 1960, the government of the Congo took the Belgian company UMHK into national ownership in 1967, placing its assets in a new state copper company, Gécamines. During the decade that followed there were widespread nationalizations across the mining sector, the timing of this activity almost certainly owing something to the buoyancy of commodity prices in the late 1960s and early 1970s and the resulting belief that there were significant rents to be had from the sector. (See copper prices in Figure 2.1.) In 1969, the governments of Zambia and Chile both took majority stakes in their copper mining industries, with Chile fully nationalizing the sector in 1971. Also in 1971, Guyana nationalized its bauxite mining industry and Mexico its copper industry. Peru took the country's principal copper producer, Cerro de Pasco, into state ownership in 1974 while Jamaica partially nationalized its bauxite mines in 1975 (Radetzki 1985). Foreign-owned iron ore producers were nationalized in Venezuela, Mauritania, Peru and Chile during the first half of the same decade (Crowson

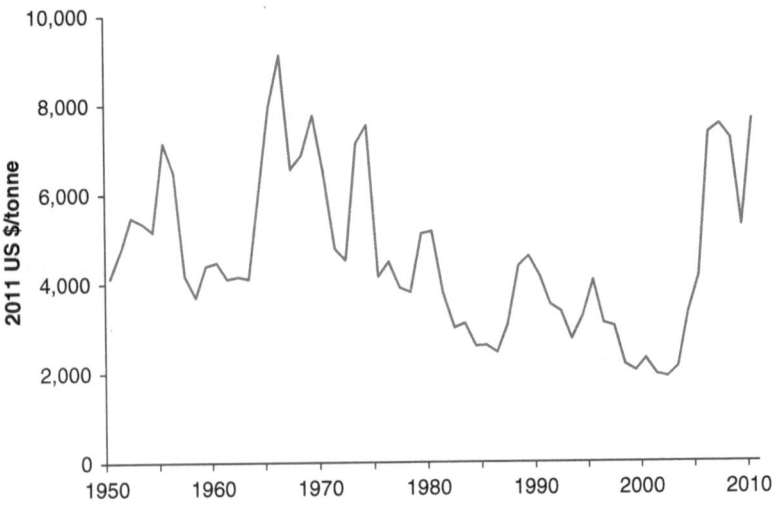

Figure 2.1   Copper price since 1950 in real US dollar terms
Source: London Metal Exchange and Bureau of Economic Analysis, US Dept of Commerce.

2008). This did enormous damage to the confidence and business interests of the mining TNCs, and some, such as the large US miner, Anaconda, which saw its assets in both Chile and Mexico nationalized, never recovered.

Important as this shift towards state ownership in the mining sector was, it did not go anything like as far as it did in the oil sector (see Chapter 1). This is was partly because the mining industry, as was pointed out earlier, is much more diverse and geographically dispersed than is oil. It is also an economically smaller industry with narrower margins and as such lacks the same 'strategic' quality of oil as well as its capacity to generate large rents. Finally, it is relevant that, despite the growing importance of mine production in developing countries, a large part of the global mining industry was still focused in countries with economically liberal traditions such as US, Canada, Australia and South Africa. This was not the case with oil where the big resources were concentrated in the Middle East.

Influence of the Cold War

Overlaying these developments in the mining sector was the Cold War and the associated polarization of the world into two ideological camps, the one led by the US, the other by the Soviet Union. For countries seeking to distance themselves from their former colonial masters, the Soviet Union's commitment to state ownership and a planned economy chimed well with governments suspicious of capitalism and eager to take control of their resource sectors. The Soviet Union, through its control over domestic resources and those of its satellite states, was largely self-sufficient in minerals, but was happy to capitalize on the disenchantment felt by these countries towards their former masters. To stem the spread of communism in these countries, and to prevent their resources coming under Soviet control, the US and its Western allies found themselves supporting some autocratic and corrupt regimes. Thus, President Seso Seko Mobutu of the Congo was able to attract the support of Western states by presenting himself as the only leader strong enough to hold the Congo together and to keep its valuable mineral resources out of Soviet hands.

South Africa represented a particular focus of concern for the forces of Western capitalism with respect to its mineral resources. By a quirk of nature, the Soviet Union and South Africa between them dominated world supply for a number of critical mineral commodities, including chromium, manganese, vanadium and platinum. These were deemed industrially and militarily strategic by Western governments. At the same time, South Africa's apartheid regime was not only hugely unpopular internationally but mounting internal opposition and unrest threatened to lead to political instability in the country and in the region. The Soviet Union sided with the anti-apartheid movement and assisted, for example, by providing training for ANC (African National Congress) leaders.

Adding to the region's political risks, two former Portuguese colonies to either side of South Africa had moved into the Soviet sphere. Following a coup in 1974, Mozambique became formally independent of Portugal in 1975. The incoming FRELIMO (Liberation Front of Mozambique) government established a one-party Socialist state and aligned itself with the Soviet Union and Cuba, from both of which it received substantial aid. To the west of the continent, something similar was occurring in Angola. Having also gained its independence from Portugal in 1975, leftist MPLA (Popular Movement for the Liberation of Angola) forces, supported by troops from Cuba, took control of the country. Despite facing almost continual civil war with Western-backed forces (notably those of UNITA (National Union for the Total Independence of Angola), the MPLA retained control of the country through to 1991 when it was formally elected to government in UN-monitored elections.

The combination of these political developments in southern Africa and a strong run-up in mineral prices over 1978–80 prompted serious concern among Western governments about possible disruptions to their mineral supplies. At the extreme, there were allegations from the US political right that the Soviet Union was engaged in a 'resource war' on the West. Although the evidence for this was never very strong and although the economic slump of 1981–2 substantially alleviated concerns over supply availability, the US, the EU and several EU member states conducted extensive investigations into these matters and some instigated programmes for strategic stockpiling of the commodities deemed most at risk (Miller et al. 1980; Council on Economics and National Security 1981; House of Lords 1982; Maull 1986) The USA's determination to have the right to mine minerals from the seabed in international waters, should its strategic interests so require, was a key factor leading to its refusal to sign the UN Law of the Sea Convention in 1981.

The north–south divide

The 'politicization' of minerals which occurred during the Cold War, both among producing and consuming countries, was not, however, very obviously delivering positive benefits for mining nations. The notion that a change of ownership was all that was required to ensure the sector would serve social objectives and promote economic development was always unduly optimistic. Disillusion was setting in and deeper concerns about the impact of mining on the political economy of developing countries were beginning to emerge.

Too often, and particularly in Africa, the resource industries which had been nationalized were not being used to benefit the people but were serving to enrich a governing elite with strong links to foreign interests. 'African leaders', observed President Kwame Nkrumah of Ghana, 'became the policemen of the western interests' and were rewarded accordingly (Ochola 1975).

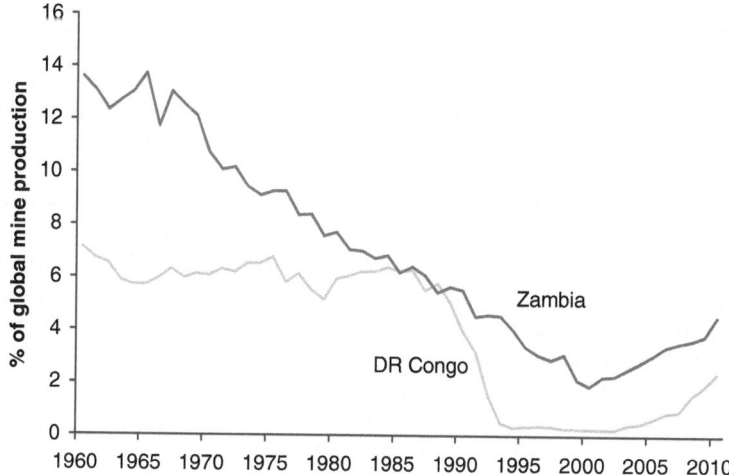

Figure 2.2   Copper mine production in Africa
Source: Metallgesellschaft and World Bureau of Metal Statistics.

Meanwhile, with sources of commercial finance cut off, the capital base of the mining industries in these countries shrivelled up and production went into severe decline. Figure 2.2 shows the decline in copper mine production in Zambia and what is now the DR Congo.

The view was taking hold that a country's possession of minerals, far from assisting its development, had the potential to actually inhibit it by distorting the economy and its exchange rate, by exacerbating inequalities, and by fostering corruption and repression (a notion which later became encapsulated in the notion of the 'resource curse'). In some cases, where the resources were located in regions with an ethnic identity which was different from that of the ruling group, such centralized state control could give rise to pressures for political secession, as occurred, for example, in the Katanga region of the Congo (then called by its Swahili name, the Shaba province) where there was an armed uprising in 1977–8, and on Bougainville Island in Papua New Guinea during the 1970s and 1980s, where a revolt by local Melanesian groups resulted in the closure of the Bougainville copper mine in 1989.

The other emerging realization was that a simple change of ownership did not result in any increase in producer pricing power. Not only were governments with nationalized mining industries having to grapple with the effects of volatile commodity prices on their fiscal flows but, worse, there was growing evidence of a systematic bias in the international trading system against commodity producers. Prices of commodities relative to those of manufactures – what is known by economists as their 'terms of trade' – appeared to show a persistent long-run decline. Something of this can be seen in the downward trend in real copper prices following the copper price boom of the late 1960s and early 1970s in Figure 2.1.

UNCTAD (United Nations Conference on Trade and Development) played a prominent role in promulgating this case on behalf of developing countries and in promoting schemes for commodity price stabilization. It also championed the downstream integration of mining countries to help them break out of the trap into which their commodity dependence appeared to have placed them. Perhaps not surprisingly, there was also much talk at the time among developing countries of seeking to follow OPEC's example in oil and form producer cartels as a means to force up mineral prices. While associations of producers and exporters were set up during late 1960s and early 1970s for all kinds of commodities including copper, iron ore, bauxite, phosphates, mercury, tungsten and silver; these never had the internal disciplines or the market coverage to give them real political leverage or to have a material impact on market prices (Crowson 2008).

At the end of the 1970s, these concerns about the conditions of commodity-producing developing countries had become widespread. An international commission chaired by former German Chancellor Willy Brandt issued an influential, and deeply pessimistic, report in which it argued that the growing wealth of the industrialized 'north' was based on the continuing impoverishment of the developing 'south' and that the two halves of the globe were on a collision course (Brandt 1980). Central to this claim was the spectre of ever-cheapening commodities. This not only condemned developing countries to remain forever 'hewers of wood and drawers of water' but would eventually lead, through starvation, environmental catastrophe and terrorism, to the general destabilization of the south and a consequent threat to the security of the north.

Picking up a theme and a label that were already current among developing and non-aligned countries, the report argued that what was required was a 'new international economic order' (NIEO) – a new global political perspective which acknowledged the mutual interests and dependence of the north and the south and a new set of international institutions to go with it. Within the commodities' sphere, it proposed a series of measures to increase the availability of finance and technical assistance to developing counties, encourage downstream integration in mining countries, and progress the commodity price stabilization and support schemes that UNCTAD had been advocating.

1980s and 1990: The triumph of economic liberalism

The Brandt report represented the high-water mark of seeking government-engineered solutions to the world's development challenges. Although the prices of many commodities remained severely depressed through much of the 1980s, the open conflict between the north and south that Brandt warned about did not materialize. Indeed, developing countries began to emerge during this period as a growing force in the world economy and a growing force in global mining. A new spirit was abroad, the spirit of the market and of economic liberalism, bringing with it a new set of recipes for economic growth and development. This change in the spirit of the times was most prominently illustrated in the West by the election, on stridently free market platforms, of Prime Minister Margaret Thatcher in the UK and President Ronald Reagan in the US, in 1979 and 1980 respectively.

However, this was far from being a Western affair. In the late 1970s, following the death of Mao Zedong, a new leadership of the Chinese Communist Party was emerging under Deng Xiaoping. This new leadership was pragmatic and modernizing, repudiating the Cultural Revolution and exhorting the Chinese people to enrich themselves. The communist regime in the Soviet Union was continuing to hang on, but it was rapidly losing its global influence as its economic power waned. By the end of the 1980s, with the liberation of Eastern Europe from Soviet control, it had effectively lost the fight.

This new wave of economic liberalism had powerful resonance within the mining sector. On the one hand, weak mineral prices in the 1980s were persuading countries that there was nothing particularly special – and certainly nothing particularly profitable – about the resources sector, it was just a business like any other. On the other, it was apparent that state ownership had not delivered the results it promised. Far from providing major economic and social benefits, many nationalized companies were highly indebted and had become a burden on the state. Gradually, and in some cases reluctantly and under pressure from lending institutions like the IMF, many countries were drawn to the conclusion that the way forward lay in seeking to attract private investment to the sector. In arriving at this conclusion they may have been encouraged by the increasing disillusion with mining in industrialized countries (stemming from high costs and growing environmental pressures on the sector) and a realization that this was strengthening the hand of developing nations in their negotiations with foreign investors on the development of their resources.

Whatever their motives, many countries set about revising their mining and taxation regimes with a view to making themselves more attractive to foreign investors. In the process, the World Bank played a prominent role, both in cajoling recipients of Bank funding to adopt more market-friendly policies and in providing significant technical and legal assistance to those undertaking reform in developing countries.[1] Between 1985 and 1996, it is estimated that over one hundred countries had revised, or were embarked on the revision of, their mining regimes (Otto 1996). Some countries began to reverse the process of nationalization and to auction off state assets. In the UK, after facing down a bitter strike in 1984–5, and demonstrating in the process that the domestic coal industry had lost its 'strategic' status on account of the growing availability of local gas supplies and cheap imported coal, the government of Margaret Thatcher broke up and privatized the UK coal industry.

TNC–host nation relations

This reorientation of the economic environment helped hugely to revive the fortunes of the mining TNCs (although it did not, it might be noted, result in a similar transformation in the prospects of the International Oil Companies (IOCS)). Having been treated as pariahs by many resource-rich developing countries through the preceding decade, some of these same countries were now soliciting the interest of foreign miners to participate in the development of their resources. Even UNCTAD, which had hitherto been generally critical of the mining TNCs, began to adjust its message to the changing times, becoming at once more supportive of foreign direct investment and turning its attention towards the terms and conditions under which such investment could benefit the economies of host nations. A landmark investment of this period was the investment by BHP (now BHP Billiton) and RTZ (now Rio Tinto) in the large Escondida copper mine in northern Chile at the end of the 1980s. Chile had been one of the lead countries in banishing foreign investors and nationalizing its mining sector almost 20 years previously.

The new period in the life of the global mining industry was marked by growing collaboration between international miners and mineral-rich host countries. This was based both on a more equitable division of power in the relationship between host country and foreign investor than had existed in earlier eras and on a deepening consensus about the conditions required for mutually beneficial mining development – most notably, security of tenure and predictability in regulation and taxation – and the need for an equitable division of the rents from mining. Glue for this more collaborative spirit was provided by the persistently unfavourable nature of the mineral markets through much of the 1980s and 1990s and the need for all parties to mining investments to keep closely focused on basic business realities. It may also have been assisted by the recognition that, in contrast to foreign investment in the era of colonialism when overseas mining development was frequently driven by the need to supply the home country with raw materials, foreign investment in mining during this new era was very much more focused on supplying global markets, an increasing proportion of which were in developing countries.

Community conflicts

With a growing consensus between the industry and the state at a national level, more prominent fault lines of conflict began to be exposed at a community level. This was in part a natural response to concerns that a central government focus on the broader economic benefits of mine development might result in inadequate attention being paid to the interests of the people mostly directly impacted by such development. It was probably also in some cases an outlet for nationalist sentiment among those unhappy about the new spirit of liberalism. Communities in the vicinity of mines naturally wanted to ensure that they got their share of the jobs and fiscal revenues that mining investments generated. By dealing directly with mining companies, communities were often able to apply pressure both on the companies and on their own governments. Challenges to companies over the impacts of their activities could be particularly effective where they were able to tap into pre-existing conflicts associated with ethnic, religious or political divisions. Thus, for example, mining companies in Australia have long had to confront issues associated with the territorial claims and human rights of the aborigine population. Miners in the Philippines have frequently run into opposition to their activities from the Catholic Church, while Freeport-McMoRan's Grasberg mine in Indonesia has long been the focus of discontent for West Papuan separatists.

Foreign investors, by the same token, were having to become much more alert to the issues of the communities in which they operated, often investing heavily in social infrastructure, not simply as a means to head off dissent but, more positively, because companies were coming to recognize that having local communities as partners and advocates significantly strengthened their hands in negotiations with central government. Accordingly, many global mining companies began to develop or scale up programmes of community engagement to foster cooperation, involving small business partnerships, community support schemes, educational activities, health clinics, conservation programmes, sports sponsorship, and participation in regional economic development initiatives.

Prior to committing investments, many companies were undertaking exhaustive consultation exercises to tease out potential difficulties and address them up front. As mining companies had found in the past to their cost, while the establishment of mines creates groups who benefit ('haves'), they also create groups who do not benefit at all ('have nots') and who in fact are marginalized and impoverished by mine developments. Identifying these latter groups and bringing them into the dialogue before mines are committed may turn out to be critical to a mine's longer-term success. Such consultation exercises commonly take several years to complete but in the context of mines which may operate for 30 or 40 years this may be time well spent. In short, companies were finding that having policies which proactively addressed the concerns of local communities could create stable collaborative relationships which served both their business interests and the economic and social interests of the communities in which they were operating.

Environmental concerns

In addition to these economic and social issues, the other factor prompting conflict at a community level was a growing concern over the environmental impact of mining. Unlike sectors where the environmental impacts are widely dispersed (e.g. emissions from cars and power stations) the environmental impacts of mining tend by their nature to be locally intense. Such impacts vary considerably according to the type of mining involved. Underground mining, although it can bring with it problems of subsidence, generally has a relatively small footprint at the surface. Strip mining for bedded minerals like coal, bauxite, nickel oxides and titanium beach sands, despite its rather sinister name, is actually one of the more environmentally benign forms of mining since it is generally superficial and permits almost complete rehabilitation of the ground and its natural vegetation after the mineral has been removed. Other forms of mining pose bigger problems. Open-pit mining creates large holes in the ground and waste dumps which are unsightly; the exposure of sulphide rocks to the elements can result in acid rock drainage into ground water; the disposal of effluent and particulate matter from concentrators can find its way into rivers and the sea; cyanide used in gold recovery carries the risk of toxicity; while mines and associated infrastructure and equipment can represent a hazard for migrating animals.

While such concerns over the environment were not new, it is probably fair to say that public awareness of these matters had grown persistently since the early 1970s when such issues were brought to public attention by such publications as *Blueprint for Survival* and *Only One Earth* (*The Ecologist* 1972; Ward and Dubos 1973). Non-Government Organisations (NGOs) were playing an increasingly active role in ensuring that local communities exposed to mining were made aware of its environmental consequences, a role that was greatly facilitated by developments in information and communication technologies such as the internet and mobile telephony. As with the case of their economic and social impacts, the mining industry (or at any rate the larger companies within it) was working to raise its standards through research and on-the-ground activities so as to increase the public acceptance of its operations. The websites of mining TNCs were filling up with statements about their commitments to good environmental practice along with data and illustrations on their work in this area. Because of the nature of the industry, however, and the intense feelings to which these matters give rise, this remains a major fault line for conflict between miners and the communities in which they operate.

There are numerous examples of environmental flashpoints, some of the most high profile involving mine tailings, a particularly challenging issue for the industry. (Mine tailings are slurries containing the waste particles which result from the grinding up of metal ores as they are processed into metal concentrates.) Thus the river dumping of tailings by the Grasberg mine in Indonesia and by the Ok Tedi mine in Papua New Guinea have been a persistent source of friction locally, as has the marine dumping of tailings at the Lihir Gold Mine in PNG and the Batu Hijau mine in Indonesia. Dam breaks at the Omai mine in Guyana in 1995, at the Marcopper operations in the Philippines in 1996, and at the Baia Mare mine in Romania in

2000, provoked loud public outcry against the mining companies concerned (while drawing wider public attention to the environmental risks of mining), the last of these having the additional complication that the polluted water flowed across the border into one of Hungary's largest rivers, the River Tisza.

Artisanal mining and conflict diamonds

The other important context in which mining activities were leading to local conflicts during this period was in the unregulated sector of informal, artisanal mining. While such mining has always existed, the high gold prices of the early 1980s gave rise to a rapid expansion of the sector, bringing it to public prominence. In Brazil, it is estimated that there were some 350,000 *garimpeiros* operating in the informal gold sector in these years, this against less than 9000 employees in the formal mining sector (Hanai 1998). These primitive, pick and shovel operations, working shallow alluvial ores, were not only damaging environmentally (not least because of their use of mercury for the recovery of the gold) but they also brought with them conflicts with the indigenous peoples in the areas they were operating. Particularly affected were the Yanomami people, who inhabited the region north of the state of Roraima in Amazonia, whose culture and health were badly impacted by *garimpo* activities. Concerned at the social and environmental consequences of such mining, the Brazilian government at the end of the decade moved to regularize the sector, formally recognizing the claims of the *garimpos* within the country's constitution in exchange for making them subject to environmental legislation.

Much more problematic were the activities of informal, artisanal miners in countries where governments lacked the capacity to regularize these activities or where the mining activities were being used to fund opposition to elected governments. This was a widespread problem in Africa, particularly with respect to the mining of diamonds. Revenues from diamonds helped sustain the military activities of UNITA in Angola even after the elections confirming the legitimacy of the MPLA government of 1991. This led in 1998 to the UN adopting a resolution prohibiting the import of diamonds from Angola which were not officially certified by the Angolan government. Sales of diamonds were also used to support the RUF (Revolutionary United Front) in their civil war against the incumbent government of Sierra Leone. In 2000, the UN passed a resolution on Sierra Leone similar to that for Angola. Other countries in Africa involved in the sale of diamonds for the funding for military activities and the purchase of weaponry were Liberia, Côte d'Ivoire, DR Congo and Republic of Congo. In December 2000, the UN General Assembly unanimously adopted a resolution on the role of diamonds in fuelling conflict, with a view to breaking the link between the illicit transaction of rough diamonds and armed conflict, and as a contribution to the prevention and settlement of conflict. At the same time, the UN acknowledged the role that the legitimate production and sale of diamonds could make to prosperity and development. Also in 2000, the global diamond industry committed itself to the introduction of a universal diamond certification process to squeeze out what had become known as 'conflict' (or 'blood') diamonds. The certification process, termed the Kimberley Process, became effective in 2003.

Alongside the enormous scale of the global mining industry, the economic significance of conflict diamonds is small indeed. Even within the diamonds sector, the trade in conflict diamonds is estimated to represent significantly less than 1 per cent of the value of all diamonds sold. However, the glamorous nature of the product and the particular brutality of the conflicts which the sale of conflict diamonds have funded, have given the issue a symbolic significance far in excess of its economic importance. Unquestionably, the negative image of the mining industry to which the issue has given rise in the public mind has gone well beyond Africa and beyond diamonds. Revenues from the sale of columbite-tantalite (coltan), cobalt, tin, tungsten and gold have all been the subject of concern about abuses in the DR Congo (Melcher et al. 2008). In July 2010, a legal action was launched in the UK courts against the government for not intervening to prevent UK-based companies producing a range of minerals in rebel-controlled areas of DR Congo (BBC 2010). The 2010 Dodd-Frank Act in the US requires companies using metals produced in DR Congo to be able to disclose the exact provenance of their supplies so as to ensure that they are not coming from illegal sources (*The Economist* 2011).

The issue of conflict diamonds also serves to make rather starkly a critical point which does indeed go across all minerals and all regions, which is that good governance is always and everywhere critical to the operation of a peaceful, productive and socially responsible mining industry, based on mutual trust and willing collaboration. A key observation of the multi-stakeholder research project, Mining, Minerals and Sustainable Development, was that in the absence of national authorities with clear public legitimacy and applying sound principles of governance, the potential for conflict to arise from mining will always exist (MMSD 2002).

2000s: Emerging economies and the return of resource nationalism

The world into which the mining industry is moving in the twenty-first century is only now taking shape and any conclusions about where the fault lines of conflict will appear must necessarily be somewhat tentative. Some of the characteristics of this world are nevertheless already apparent. The 'new normal', as it has come to be known, is a world in which emerging market countries will play a prominent and expanding part in the global economy, and thus its polity. Moreover, while capitalism may have triumphed as a result of the ending of the Cold War, the model coming to prominence in

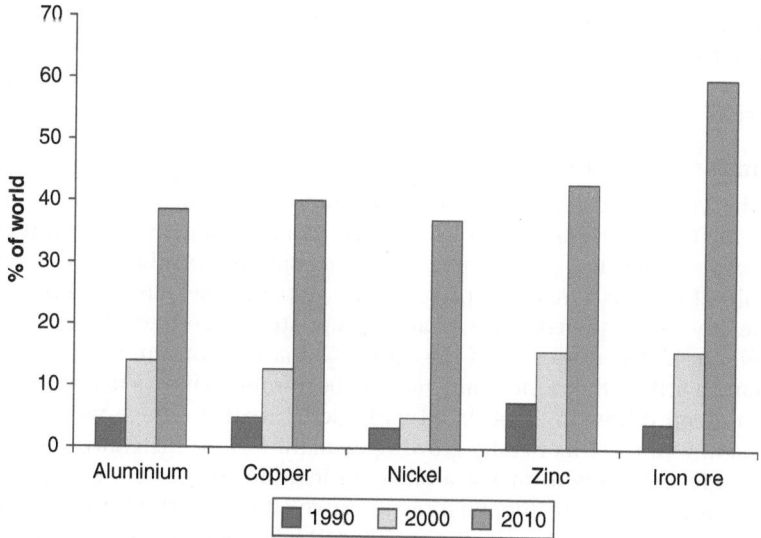

Figure 2.3 China's share of global mineral demand
Source: WBMS, UNCTAD. Iron ore data relate to seaborne iron ore.

this new century is not that of market capitalism but, increasingly, that of state capitalism.

The rapid economic growth of China is re-shaping global economic and political relationships while the upsurge in the demand for mineral products needed to fuel its industrialization has helped transform the fortunes and prospects of the mining sector. As illustrated in Figure 2.3, China's share of global metal markets grew from under 5 per cent in 1990 to around 40 per cent in 2010 (and nearer 60 per cent in the case of seaborne iron ore). It might be noted that this was not a situation that was supposed to occur according to the world view at the time of the Brandt Commission when it was considered axiomatic that the countries of the north would always provide the primary dynamic for mineral demand.

Demand from China was a key factor helping to drive the commodity price boom of 2003–08, a boom which effectively marked a transition to a new and distinct phase in the life of the minerals industry, much as the price boom of the late 1960s and early 1970s had marked a similar transition 30 years before (Humphreys 2010). As with the earlier boom, higher mineral prices, combined with the conviction that these would have a sustained impact on mineral asset valuations (there was much talk of a mineral 'super cycle'), have triggered a new round of rent-seeking behaviour. Specifically, it has stimulated a renewed interest from the state in the mining sector and led to the state's increased involvement in its affairs worldwide. If growth in demand from the emerging economies is the first key characteristic of this new chapter in the life of the industry then the growing role of the state is the second. These two tendencies potentially bring with them a whole new range of conflicts and, perhaps, the resuscitation of some old ones.

China's 'go out' policy

In the years following its economic liberation by Deng Xiaoping, China was for the most part content to source its raw materials from domestic resources which it supplemented by buying supplies on the open market. While China was a marginal player in the global market, and markets were generally well supplied, this was an appropriate and cost-effective strategy. However, by the early part of the 2000s, China was consuming something around a fifth of the world's minerals and beginning to have a material effect on mineral prices. In 2004, faced with global markets which were clearly struggling to meet its rapidly growing demand for raw materials, and concerned lest this might impose a brake on its industrialization, the Chinese government started to adopt a more assertive attitude with respect to securing its supplies. In this year, it promulgated a 'go out' policy, exhorting its large companies (many of them state-owned) to seek out and secure supplies for home consumption by investing in natural resource projects overseas. In 2007, it added force to this mission by establishing a sovereign wealth fund, China Investment Corporation (CIC), with an endowment of $300 billion for foreign direct investment, which was to include investment in minerals.

With commodity prices hitting record highs, and few mineral assets available at reasonable cost, progress in this mission was initially slow. The first investments by Chinese companies were early-stage development projects and minority stakes in foreign mining operations made in return for supply off-take agreements, rather than outright purchases of existing mining companies. This reticence towards outright acquisition was doubtless in part attributable to the political storm unleashed in Canada by the attempt of China Minmetals to acquire Canadian miner Noranda in 2004.

Following the downturn in commodity prices during the second half of 2008, which not only reduced asset values but also made companies more open to the approach of new investors, Chinese companies stepped up the pace of their buying. Thus, for example, Chinese aluminium producer Chinalco acquired 9 per cent of Rio Tinto in 2008, while in 2009 China Minmetals acquired the assets of Australian miner OZ Minerals and Hunan Valin acquired 18 per cent of Australian iron ore producer Fortescue Metals Group and CIC acquired 17 per cent of Canada's Teck Resources (Humphreys 2009). Nevertheless, significant opposition remains to the expansion of China's participation in the sector in this way and all has not gone smoothly. Minmetals was debarred by Australia's Foreign Investment Review Board (FIRB) from acquiring the Prominent Hill copper mine as part of its bid for OZ Minerals, while FIRB effectively blocked the attempt of the China

Nonferrous Metals Corporation (CNMC) to acquire the Australian rare earths producer, Lynas Corp. Opposition from Rio Tinto's other investors played a part in preventing Chinalco from increasing its stake in the company during 2009. Acquisitions by Chinese state-owned companies will doubtless continue in the future and this will remain a source of latent conflict in Western countries.

As in the case of oil, China's quest for mineral supplies has also been leading it deeply into Africa, giving rise to a rather different range of stresses. China's approach to mining investment in Africa has shown marked differences with that of Western TNCs. While Western companies have shown extreme caution in investing in the region because of the political and environmental challenges they confront in Africa and the associated risk to shareholders funds, Chinese companies have shown little fear or restraint. Such companies are typically much more focused on the long-term strategic nature of investment and many of them, being state-owned, are not constrained in the same way by the concerns of private investors. Their investments are as much about securing supplies as they are about profits.

For the most part, these investments have been welcomed by host governments. In part this is because China's investments have often come with significant additional development aid for infrastructure and power. It is probably also because China's principle of non-interference in the affairs of other countries makes it less judgemental about the nature of the regimes it is dealing with and less likely to demand tough conditions in exchange for its investment. However, while governments can do deals with companies, this does not always mean they bring their people with them, as the Western TNCs found in the days of neo-colonialism, and the Chinese tendency to import large numbers of their nationals to staff their operations in Africa has created significant tensions at a local level. Press reports talk of frequent clashes between Chinese and local populations (*The Economist* 2006), while the presence of large numbers of Chinese in the country was a prominent issue in Zambia's elections in 2006, 2008 and 2011. More recent mining development agreements between the Chinese and the government of the DR Congo have had written into them clauses restricting the number of Chinese workers permitted to work on the projects (Komesaroff 2008). Chinese investments in Latin America have similarly run into local resistance. For example, Zijin Mining has faced vigorous opposition from local groups to the development of the Rio Blanco copper mine in Peru.

Resurgence of the state

The growing participation of the state in the global mining industry is by no means an exclusively Chinese phenomenon. The soaring profits of the global mining companies in the mid 2000s led the governments of many mining countries to the view that they were receiving an inadequate share of mining rents. This view was bolstered by a widespread conviction (encouraged by the mining industry itself) that commodity prices were likely to stay at elevated levels for many years to come and, in some quarters, by a belief that minerals were getting scarcer. Many countries which had established their mining regimes during harder times, in the 1980s and 1990s, with a view to attract foreign investment, determined that they had given too much away and revised their mineral royalties and taxes upwards in response to these new, more buoyant market conditions. These countries included Chile, Peru, Zambia, Tanzania, South Africa, Indonesia and Vietnam (*The Economist* 2007). There was also a revival of interest in the notion of nationalization, particularly in Latin America, with Venezuela, Bolivia and El Salvador all taking mining operations back into state ownership or revoking mining permits previously awarded to foreign mining companies (Lozano 2009). In 2007, the government of Zimbabwe passed legislation aimed at indigenizing control of certain key sectors, including mining, by requiring them to be 51 per cent owned by local interests. The legislation came into force in March 2010.

Another aspect of this growing resource nationalism is the increasing prominence on the global mining scene of mining companies from the emerging market economies. While these are not necessarily state-owned or state-run companies, they are nevertheless in many cases very much national companies, having grown up largely in one country and being perceived in some sense as national champions. These include such companies as the state-owned copper producer, Codelco, in Chile, and the publicly quoted Vale (formerly CVRD) in Brazil (in which the state maintains a very close interest and retains a 'golden share'). They also include a number of companies from former communist countries such as Norilsk Nickel and RusAl from Russia, Kazakhmys and ENRC from Kazakhstan (in both of which the state holds a minority interest) and KGHM Polska Miedz from Poland. Benefiting from a strong local resource base and new-found access to global capital markets, these companies have grown strongly in recent years (Humphreys 2011).

The emergence onto the global stage of such state capitalist enterprises from China and elsewhere is posing a major challenge for Western mining TNCs. Having devoted enormous efforts over recent years to the issue of Corporate Social Responsibility (CSR) and to upgrading their environmental performance with a view to positioning themselves as 'developers of choice', they find themselves in some parts of the world competing with companies from elsewhere on wholly different grounds to those they had expected. Despite the tidal wave of mining reforms in the 1980s and 1990s, the number of countries in which these companies can invest, confident in the knowledge that their permits and their assets will be protected under the rule of law, are still quite limited, and probably shrinking.

The military regime of Guinea which came to power in December 2008 determined that all mining rights issued under the previous regime would be reviewed, including Rio Tinto's licence for the giant Simandou iron ore project and RusAl's licence for the Friguia bauxite-alumina operations. As a result of this process, Rio Tinto was stripped of two of its licences. This followed a similar initiative by the incoming government of DR Congo in early 2007 under which it pledged to review over sixty mining licences issued under preceding regimes during 1998–2005, and which resulted in several companies having their licences withdrawn. In 2010, the government of DR Congo expropriated two mines operated by Canadian miner, First Quantum Minerals. In countries featuring dominant national players, such as Russia and Kazakhstan, the inside knowledge and political contacts of the domestic players will always put outsiders at a competitive disadvantage, requiring them, if they are to invest in these countries at all, to do so in partnership with a local enterprise.

In short, the industry is moving into a world where business relationships will have a less legalistic and a more political foundation. The world is not going to be reshaped by the triumph of liberal democracy as some believed at the end of the Cold War. Weak application of law along with uncertainties over taxation will likely create many flashpoints for conflict between companies and governments in coming years. Companies will increasingly find themselves having to deal with governments and to negotiate the terms of their investments on an *ad hoc* basis.

The lengthy negotiations between Ivanhoe Mining and Rio Tinto on the one side and the government of Mongolia on the other over the development of the large Oyu Tolgoi copper mine (officially concluded in 2009 but which the government sought to re-open in 2011) may represent the pattern of things to come. If so, then it is likely that the mining TNCs will increasingly have to look to their own governments to bolster their positions in negotiations and provide them with quasi official status. In this process, it may also be that lines between what is politics and what is business will become increasingly blurred. The arrest of four Rio Tinto employees by the Chinese state authorities on a charge of industrial espionage in 2009 (later changed to a charge of paying bribes) helped highlight the potential problems that can arise when the lines between business and state activities lack clarity. Potentially, the mining TNCs are looking at a future not unlike that of the IOCs, ruled out of operating in certain mineral-rich regions by political and legal uncertainties and forced to compete in others on disadvantageous terms. Viewed against this backdrop, the wave of mergers and acquisitions among Western TNCs during 2004–8, which revealed a preference among these companies to grow by acquisition of like-minded companies rather than by undertaking high risk new investment, is more easily understood. The IOCs had undergone a similar consolidation in the late 1990s.

Current directions

As emphasized in Chapter 1, the conditions facing natural resource industries have a cyclical quality and the world into which the mining industry is
moving has features of previous regimes and some novel features of its own.
It is multipolar and multilayered. It lacks the simple bi-polarity of the Cold
War years and the unifying influence which was provided by US hegemony
(and the US dollar) in the post-Cold War era. At the same time, decreasing central government authority in many countries, combined with the
radical effects of improved communications technologies, have empowered
local communities across the globe. The potential for conflict in the mining
sector within this complex and volatile world is considerable. By the same
token, the rewards for understanding what is going on and building models
of collaboration to address nascent conflict are also potentially large.

Goldman Sachs, in its research on commodity markets, has employed the
notion of 'terra concentration', a term signifying the amount of the earth's
surface falling under the control of one of the great powers of the day (Currie
et al. 2008). They point out that the world's natural resources have not been
controlled by so many separate political entities since the seventeenth century when mercantilism was the ruling economic ideology. The notion is
shown graphically in Figure 2.4.

This fragmentation of control need not be a problem for mineral supply if capital is free to move around the globe. But this is not the case,
and the barriers to such movement are increasing. Goldman Sachs is led

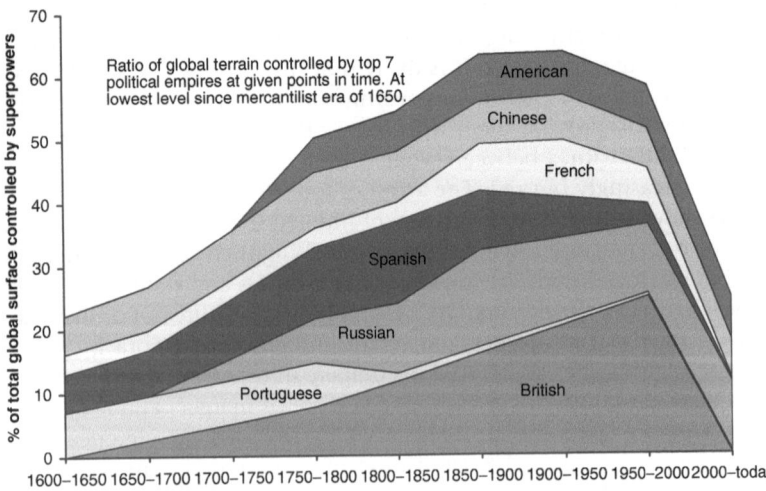

Figure 2.4   Terra concentration
Source: Currie et al. (2008).

to conclude that the fragmentation of control over resources and the protectionism of resource nationalism are resulting in inadequate and inefficient industry investment, as a consequence of which supply will be constrained below where it would otherwise be and commodity prices will remain under strong upward pressure for years to come. The problem, as Goldman sees it, 'is about the supply of capital, not the supply of the commodity'. While Goldman Sachs's thinking on this matter is strongly shaped by the oil sector, where resources tend to be more geographically concentrated than in the case of minerals, the argument nevertheless is highly relevant to the minerals sector.

Linked with the fragmentation of control over global supplies is another issue, which is the growing distance between producers and consumers of minerals. In the world of colonial powers, both the production and use of minerals tended to fall under control of the same set of countries. Even in the post-Second World War era of economic dominance by the OECD countries on the one hand and the Soviet Bloc on the other, it was still generally the case that the same group of countries dominated both the production and consumption of minerals. In these circumstances, there was a clear self-interest in mineral-consuming countries to devise and implement policies which encouraged investment in mine development.

However, recent years have seen the extent of this overlap between producers and consumers diminish (a development facilitated by advances in bulk transportation), progressively removing this element of consumer self-interest from the process. Much of the growth in demand for minerals in these years has come from East and Southeast Asia, most notably China, South Korea, Taiwan, Malaysia and Thailand, countries whose limited resource endowments make them natural importers of minerals.[2] At the same time, much of the supply growth in minerals has come from emerging market countries with quite limited domestic demand, for example, Chile, Brazil, Russia, Central Asia, Indonesia and southern Africa, and from the large developed world net exporters, Australia and Canada.

These trends are increasing the amount of global mineral supply which is subject to international trade. They are also changing the dynamic of producer–consumer relationships and placing producers and consumers of minerals into two separate camps with somewhat opposed interests. While in principle the market should be able to mediate efficiently between the two groups and provide economic benefits to both, in practice, for producers without significant resource-consuming interests, the principal concern in an era of state capitalism will tend to be the profitability of their activities rather than ensuring abundant low-cost supplies to consumers and this goal may be as well served by restricting supply and keeping prices higher. (Once again, the parallel with the oil industry and the activities of OPEC come to mind.) Within the confines of the WTO (World Trade Organisation), the long-established principle of a country's sovereign right to dispose of its natural resources as it sees fit is increasingly being used to challenge the notion that all buyers within the WTO should have equal access to minerals once mined, as illustrated, for example, by China's controversial imposition of limits on its exports of rare earth elements (Milmo 2010). At the same time, attempts by consuming countries to reduce their dependence on mineral imports, or else to seek to import unprocessed minerals for home processing as a means to enhance the security of supplies of raw materials for their industries, risks depriving mineral producing countries which want to develop their resources of the opportunity to use these resources as levers for their own economic development. The combination of these pressures potentially sets up mercantilist-style nation-based conflicts between those countries with the resources and those that use the products from them. In this increasingly politicized environment, countries with a strong statist backing for their enterprises are better equipped than those without to prosper.

Response of importing nations

The unfolding of this tendency has led to renewed concerns over the security of supply of minerals in the US and Europe, just as happened following the commodity price booms of the 1970s. Whereas the concerns last time around were over South Africa and the Soviet Union, this time the focus of concern is China and the fear that its buying power, its restrictions on certain commodity exports and its investment in tied mines overseas might leave other consuming regions struggling to obtain supplies. There are differences also in the commodities which are the focus of concern. The growth in the importance of electronics to industry and to the military during the 30 years since this was last a major agenda item, has thrown the spotlight onto the so-called electronic metals like indium, tantalum, ruthenium, germanium and, above all, the rare earth metals, for which China dominates global supply. However, while these matters have been the subject of numerous resolutions and reports (e.g. European Commission 2010; National Research Council 2008), and have been extensively debated in political forums, thus far they have not resulted in any major policy action, this perhaps reflecting a lack of political will or a hope that the market will eventually resolve the supply problem without the need for intervention.

Another focus of concern which has re-emerged in consuming countries in recent years relates more to the issue of the physical availability of minerals rather than geopolitical availability. One manifestation of this thinking is encapsulated in the notion of 'peak minerals' (see Chapter 11), an idea which echoes the concerns of the Club of Rome in the early 1970s (Meadows et al. 1974) that resources may be 'running out'. Another, rather more plausible variant of this thinking is the concern that the increasing cost of producing mineral raw materials, combined with greater project and currency risk, may obstruct the timely development of new resources. There are strong grounds for believing that, for many mineral commodities, the more accessible, lower cost, resources have already been exploited and that in future miners are going to have to turn to smaller, deeper, lower grade or more remote deposits. This will involve higher capital expenditures and higher operating costs. Companies will have to be persuaded that they can manage the risks involved in taking these projects on and that commodity prices will be sufficiently high to warrant them doing so. Although there are many who believe that prices have moved onto a higher plane since the boom of 2003–8 and that the long price downtrend since the 1970s has been reversed, there is still much debate about the scale of such price increases and it could yet prove that new supplies come through at a slower rate than mineral consumers would like.

Pressure from mineral uses

A final potential fault line for conflict over minerals relates not to how or where minerals are produced but to how they are used. This has not on the whole been a major issue in the past. Environmental NGOs have long protested about the environmental impacts of gold mining on the grounds that gold serves no real social purpose, but such criticism has hitherto rarely had any material impact on real-world decisions. Uranium has also been a focus of concern on account of its radioactive nature and the consequent risks posed by its recovery, transportation and processing as well, of course, as its potential uses in the production of nuclear weapons. Because of these risks, the industry's operations are monitored on behalf of the global community by the International Atomic Energy Agency (IAEA) while the Nuclear Non-Proliferation Treaty of 1970 exists to limit the spread of nuclear weapons. However, the refusal of Iran to cooperate with the global community on these matters well illustrates the limitations of these powers when they are challenged by states which believe they conflict with their national interests and sovereign rights. The renewed interest in nuclear power because of climate change is leading to an increase in uranium production and will presumably aggravate this problem further.

For the future, the use of coal may come more into the spotlight because of its alleged role in climate change. Coal emits more carbon per unit of heat delivered than oil and gas. On the other hand, it is widely available throughout the world and plays a major part in power generation in the two emerging industrial giants, China and India. These countries have limited power generating options so the use of coal will not be easily or quickly curtailed. Accordingly, while in the past the primary focus of concern and of conflict for the mining industry has tended to be on the upstream issues of access to minerals and the impacts of mineral recovery, it is possible in the future that more attention will be directed towards the downstream issues of minerals use and the capacity of the planet to absorb the impact of the continued growth in use. As the debate over climate change well illustrates, this has the potential to stir up a major north-south rift with strong echoes in history.

Conclusion

The historical analysis contained in this chapter serves to emphasize that the sources of conflict in mining move around with time. Many sources of conflict are recurring albeit that they come back each time in a slightly different form. Early conflicts in mining flowed from the essentially enforced nature of many of the relations within the industry; notably, the subjugation of workforces and colonial oppression. In the post-Second World War period, the focus of conflict shifted more towards the level of the state and was marked by inter-state conflicts over ownership and control of mineral resources. Subsequently, the focus shifted more towards the regional or community levels and towards environmental issues, including the issue of the use of minerals. In coming years, it is likely that the world will see vestiges of many, if not all, of these conflicts running concurrently. Contemporary issues identified as of particular interest in shaping the future policy context are the growing role being played in both mineral production and mineral consumption by emerging market economies, a resurgence of the role of the state and of resource nationalism in the minerals sector, and the diminishing overlap between the cast of world mineral producers and mineral consumers.

Notes

1. This liberalization of mining regimes in developing countries did not, it should be noted, translate into a large and fast redistribution of global production of minerals. This is partly because mineral production, being long-term and capital-intensive in nature, tends to be very 'sticky', but also it is because the mining regime is only one of a number of factors mining companies take into consideration when deciding whether or not to invest in a mine; political risk and mineral prospectively are obviously other important considerations. It has been pointed out that, in practice, much of the additional investment flowing to developing countries as a result of these changes was concentrated on relatively few commodities (notably copper and gold) and relatively few countries (notably Chile, Peru, Indonesia and Papua New Guinea) (Bridge 2004b).

2. China is clearly a major mining country, particularly if one takes account of coal and industrial minerals. It remains the case, however, that a very substantial proportion of its domestic production of the most economically important metals, steel, aluminium, copper, nickel and zinc, are based on imported raw materials.

3
History of the Gas Industry

Chris Cragg

The gas industry has grown from local roots to become a global industry. It is fundamentally different from oil, being a much cheaper commodity. It has been through major transitions in ownership from its private pioneering days to municipally owned local networks, then centralized under government ownerships until privatized again. Currently, while the distributors of gas in the West remain largely privately owned, the international suppliers are largely state-owned bodies in Russia and the Middle East and as such part of the trend towards state capitalism.

Also in contrast to oil, the gas business has largely grown up without significant conflict, requiring a high degree of cooperation between supplier and distributor. Partly this is the consequence of the complexity of its technology, but also because of the mutual dependency of both buyer and seller. Gas needs pipelines in a way that oil does not. Without distribution networks, the gas is essentially valueless, indeed dangerous in terms of climate change. As such lines have got longer, they can only be built and filled by mutual consent often with very large government-to-government deals. Only subsequently can a free competitive market be developed with smaller deals.

The natural gas industry now plays an important role in the provision of electricity, following the development of the gas turbine. This reinforces the importance of a continuous supply for the development of a national economy. While oil embargoes threaten transport, unless the embargo results in an overall reduction in the availability of oil worldwide, they often merely divert the flows into different channels, thus increasing costs, but not causing significant shortages. By contrast, given the nature of the infrastructure, the international trade in natural gas is rarely embargoed, for the consequences to the importing economy in terms of both heat and electricity are extremely great.

History of gas as different from the history of oil

The use of gas as a fuel is both older and younger than the use of oil. It is older in the sense that gas manufactured from coal was first used to light city streets in the early part of the nineteenth century, almost a 100 years before the first gasoline-fuelled vehicles drove along those streets. It is much younger in the sense that the industry's transformation into the enormous international business of today only really began in the late 1960s.

The history of gas is thus fundamentally different from that of oil. Crude oil, as the major source of refined petroleum products, had an international element even from its earliest days. Gas, initially derived from coke and coal, was largely a local resource. While the US can claim correctly that it was the first country to develop oil for heating and this was initially local to Pennsylvania in the mid-1840s, crude oil's value saw its sources proliferate from Romania to the Caspian and on into the Middle East extremely fast. By contrast, gas for lighting and subsequently heating derived from local manufacturing and extended only as far as pipes could be laid to carry it. Its international dimension only started with the development of liquefied natural gas (LNG) and the creation of very long-distance large diameter pipelines, principally in the former Soviet Union.

As a result of this localism, the history of the development of the gas industry has much less of the geopolitical allure of the oil business. Prior to the late 1950s the industry was not perceived as being related to the international energy business at all. Indeed, its internationalism has evolved only slowly and largely within the past 40 years. Equally, until recently it has largely been untouched by the concept of supply security, which is so overwhelming in relation to crude oil. Nor has the price of gas been seen as a major indicator of economic fortunes, except in so far as it has been pegged to oil. The reasons for this are several, the most obvious being that crude oil is a liquid and is much easier to store, or transport. It does need to be refined but even in the early years of the oil industry this was comparatively easy if imprecise, by simple distillation. Gas of any kind is wedded to its transport and a far more complex storage infrastructure.

Equally important is the fact that gas as a fuel has much less calorific value than oil products generally. This has a considerable technical and economic impact. For example, one litre of ordinary fuel oil has roughly the same calorific value as around 39 cubic feet of natural gas. Leaving aside the obvious convenience of transporting a mere litre of liquid, compared with trapping and transporting 39 cubic feet of gas, the use of gas as a fuel requires transporting a great deal more of it, than oil for the same result. It has far less energy density. Depending on its type, 1,044,000 kilolitres of crude oil has roughly the same thermal value as 1,000,000,000 cubic metres of natural gas. Since the volume of a kilolitre is equivalent to a cubic metre of liquid, crude oil has roughly ten times the heating value by volume as its gaseous rival.

The most common ingredient in both manufactured 'town' or 'coal' gas and natural gas is the simplest hydrocarbon molecule methane, consisting of four hydrogen atoms linked to one of carbon; CH_4. While this has advantages in terms of reducing the carbon dioxide created on combustion, it explains its low calorific value in relation to the long carbon/hydrogen and oxygen molecular chains that are contained in petroleum products. Thus, in so far as the main ingredient of fuel gas is methane, it is a clean-burning fuel that burns with a clear blue flame producing roughly 1000 British thermal units for every cubic foot consumed. To further differentiate the two fuels, the gas industry did not originate from drilling rigs at all, but from the gasification of coal and coke, the product being named variously 'town' or 'coal' gas. In essence, what is discussed here is the use of mostly methane shipped through distribution grids.

For the purposes of clarity, it is necessary to clear up a few confusions. For methane is not the only gaseous hydrocarbon at atmospheric pressure. Next in hydrocarbon chain of molecules come three other gases: ethane (C_2H_4), propane (C_3H_8) and butane (C_4H_{10}). All three of these have a higher calorific content than methane and may also be found in small quantities in natural gas. The higher the proportion of these other gases in the natural gas, the higher will be the thermal content. The main role of ethane is as a valuable petrochemical feedstock, while propane and butane are more familiar under pressure in the form of liquid petroleum gas or LPG. This latter is most familiar as the 'camping gas' used for cooking worldwide and increasingly as a gasoline substitute in vehicles. While performing a valuable role within the gas industry, LPG is not to be confused with liquid natural gas or LNG, which is largely methane cooled into liquid form, or indeed with NGLs, which are natural gas liquids, like natural gasoline, which emerge from the heat and high pressure of the ground as gases, but are liquids at normal atmospheric pressure and temperature.

Early days: Birth of an industry

The first recorded use of gas down some kind of pipe for heating and lighting comes from China in the third century BC, although it did not appear to have been widespread. This showed ingenuity at a time when our superstitious European ancestors were scaring themselves over the occasional sighting of 'will o'the wisp' blue flames on the surface of ponds. Later still Marco Polo reported seeing mysterious blue flames emerging from rocks near Baku on the Caspian but does not record whether they had any function other than curiosity value. Persian emperors apparently used this naturally occurring gas for cooking, but only in situ. Right up until the eighteenth century, the primary sources of heating remained coal and wood, while illumination was provided by candles, animal fat and less obviously oils from whale blubber. A reliable source of illumination was obviously required to break out of the cycle that meant most activity was governed by the sun.

The discovery that coal could be synthesized into a flammable gas in England is widely accredited to one Rev. John Clayton sometime after 1684, but it took another century before Philippe Lebon, a Parisian chemist, patented a system for doing so in 1799. As Clayton had done in a more primitive manner, Lebon put coal under pressure and heated it producing a mixture of hydrogen, methane, carbon monoxide and ethylene. Lebon demonstrated its potential for street lighting in Paris in 1801, but the idea did not take off. It was Frederick Albert Winsor, an anglicized German previously known as Winzer, who had visited Lebon to see his device and brought the idea to London. He persuaded the city wealthy to invest in illuminating Pall Mall in 1807. The Gas Light and Coke Company was founded in 1814 and celebrated by lighting up Westminster Bridge on New Year's Eve. The gas was funnelled through wooden pipes (Barty-King 1985).

The success of town gas as a source of illumination spread across Europe and across the Atlantic at great speed. Baltimore had its first street lighting in 1814 and Fredonia in New York by 1821, although the latter – significantly – used local natural gas from a nearby marshy creek. A Dublin theatre was lit up in 1823, while the first city in Germany, Hannover, installed some street lighting in 1825. At the same time gas lighting was still a rather dull affair reliant on reflected light. It was better than candle-power, largely because of its ethylene content which burnt with a brighter flame, but the hunt was on for a material that would not combust while turning the largely blue flame into greater light. This came with the arrival of the carbon filament lamp, whose US patent arrived in 1845. From then on coal gas started to be taken seriously for illumination not merely for streets, but also for domestic homes. The arrival of gas pipes in the residential sector was underway and with it the concept of the gas ring main and the technical means of sustaining pressure in the system. By the 1860s, coal gas lighting was widespread across Europe and the coastal cities of the US, finally arriving in Osaka, Japan in 1870s. By 1878, gas lighting was sufficiently accepted that the New Opera House in Paris could be lit by some 9200 naked flames, 556 of them on a single chandelier and supplied by 25 km of pipes.

The gas was largely used for light not heat. Coal still provided the latter. In addition, given the level of available pipe technology, the enthusiasm for gas lighting was not without its risks of fire. However, this has to be put in the context of the previous technology, notably the open flame of the candle and gas illumination may well have reduced the risk of fire, rather than increased it. A further hazard, true right up until the 1960s was that the carbon monoxide in the coal gas was lethally poisonous, creating the idea – in later years – that suicide could be easily achieved by putting your head in the oven.

It needs to be emphasized that the industry's roots were local. By the 1870s, Germany had 340 largely municipally owned gas companies and by the end of the century France had over 500. At the time of the UK Gas Act in 1948, which nationalized and centralized the system, there were over a thousand privately owned or municipal gas providers, rationalized by the Act into 12 area boards. American providers were largely private and the US industry remains highly diverse, with the American Gas Association representing some 195 gas utilities to this day.

The growing gas lighting industry did face a challenge when Edison perfected the incandescent bulb in 1880. However, electricity was expensive and not widely available prior to the First World War. It was also regarded with suspicion and, with some justification given the wiring, highly dangerous. In the UK at least it was not until the Electricity Supply Act of 1926 that electricity in the home became relatively commonplace. Gas-fuelled illumination was still growing in 1920 and only disappeared completely in the 1950s (Pearson 2009). By that time, the range of uses for gas in the home, most notably for water heating and cooking had expanded enormously, as had the level of demand. Valliant produced its first iconic 'Geyser' wall-mounted gas-fired water-heating unit in 1905. Baltimore got its first gas-fired domestic central heating system in 1916 and the first AGA gas-fired cooker arrived on the scene in 1922. By the mid-1920s most households in Europe were cooking with coal gas.

Post-1945 switch to natural gas

As noted, the US was the first nation in modern times to use natural gas instead of coal gas for municipal lighting in Fredonia in 1821 and thus to use the natural resource. It was also the first country to create a long-distance pipeline of 120 miles from southern Indiana to Chicago in 1891. The US had the advantage that it had already developed a substantial oil industry, with large quantities of 'associated gas', or gas that was produced with the crude oil. Its disadvantage was that these developed oil reserves in places like Texas and Oklahoma tended to be a long way away from the cities where the demand for natural gas was highest. To produce the oil, it was necessary to flare the gas.

This did not mean that the oil industry was unaware of the potential value of the associated gas it was flaring, merely of the limitations of pipeline technology and the high capital costs involved in getting the gas to market. For a start, liquid pipelines were easier to construct. If they leaked, the contents did not completely disappear into the atmosphere. Equally, the low relative calorific and thus low monetary value of the gas content meant that higher pressures were needed and hence greater stress on the welds. The invention of electric arc welding in 1925 did help matters and there was brief burst of unregulated US gas pipeline construction, prior to the Great Depression. However, the companies which got involved tended to be highly leveraged and speculative and suffered accordingly in the economic downturn (Castaneda and Clarance 1996). As elsewhere in the world, gas pipelines have a high capital cost and a relatively low rate of return.

US gas pipelines only really began to snake across North America after the Second World War, which had also provided technical improvements and mass production techniques. It says a lot for the massive expansion of the US gas transmission network in the 1950s and 1960s that it now reaches 250,000 miles and is almost 100,000 miles longer than the oil pipeline system, reaching far into both Mexico and Canada. These transmission lines are attached to an estimated million miles of distribution mains at much lower pressures.

In western Europe, it was the discovery of North Sea oil and gas that made the switch from manufactured gas to natural gas and the leaders were the Dutch. They had first found natural gas as early as 1948, but the real discovery occurred in 1959, with the first well at Slochteren on what subsequently became the massive Groningen field. Faced with the dilemma of what to do with this massive new resource, the Dutch government created a joint venture – Gasunie – with Shell and Exxon, taking 50 per cent and giving the companies equal shares of 25 per cent in 1963. This rapidly started to supply the old coal gas distribution networks, given the obvious advantage that the new natural gas supply had a considerably higher calorific value than what it replaced (Peebles 1980).

Taking a leaf from the Dutch book, BP found the West Sole gas field in the southern North Sea in 1965 and brought it on stream two years later. This was followed by such a significantly large number of fields in the area that the British Gas Corporation, now consolidated and in effect a nationalized monopoly supplier, took the choice in 1966 to make the switch from coal to natural gas. This was no mean decision. Largely due to the higher calorific value, the burner tips of some 40 million appliances had to be changed or checked and access gained to some 13 million households. The distribution system had to be vented of coal gas and flared, while all systems were turned off before the natural gas could replace it. It cost £563 million in money of the day and took the best part of a decade.

In fact, the big switch to natural gas from manufactured gas tended to help consolidate the gas industry in Europe, not least because of the cost involved. Germany had discovered its own natural gas source at Bentheim in 1938 and by 1942 was supplying the network with 3 billion cubic metres a year (Bcm/y). However, the proportion of natural gas used in the gas supply network was still only 10 per cent of a grand total of 7 Bcm/y supplied by Ruhrgas, Germany's largest gas company by 1965. Shortly afterwards, the company made the same strategic decision as the British Gas Corporation to make the switch, but with the difference that it needed major imports to supply it It made deals with other southern German utilities in the late 1960s to expand the network southward and even created an agreement with SNAM, Italy's major gas company to re-export into Italy anything the Germans did not use (Ruhrgas 1991).

Enter the long-distance pipeline

The perceived source of this new European gas supply was the USSR and the deal was highly political. Chancellor Willy Brandt had long been seeking a potential means of reducing tension with his Ostpolitik and a new deal between Ruhrgas and Soyuzneftexport provided the means. An initial agreement was made in 1970 for the supply of 3 Bcm/y of Siberian gas to the German border, with the German steel-maker Mannesman AG providing pipe. In 1974 the contract was expanded to provide for an annual delivery of 9.5 Bcm until 2000. In the period of détente at the time, this was not seriously opposed by the US government, but a decade later, particularly under President Reagan, it became a source of friction, when the Germans agreed to fund an entirely new gas line purely devoted to exports from Western Siberia. For the first, but not the last time, Europe's gas supply dependency raised a question of geopolitical security (Pirani et al. 2009).

Soviet supplies were abundant particularly in Siberia. The problem lay in the fact that they were around 4500 km away from the German border. Nonetheless, in spite of US opposition, this first major line from Urengoy was funded by a consortium of German and other European banks to the tune of DM 3.4 billion, the steel was provided by Mannesmann and the Japanese, technical help came from the UK's John Brown and the compressors provided by Italy's Nuovo Pignone. It came on line in 1984, an astonishingly short two years after the first ground was broken.

This first truly major deal was significant for several reasons. It was the first international attempt at a gas pipeline of this length and it used large 1400 mm diameter pipe; a Russian speciality. If Western engineers threw up their hands at the seemingly gay abandon with which Russian crews welded their way up many kilometres of line per day, this was something the Russians had done before on their own network. With a total capacity of 32 Bcm/y, it was not merely longer but bigger than anything seen before outside the USSR. Secondly, according to the standard rules of engagement in Western oil and gas companies, it made no financial sense at all. No Western company or consortium would have built it, because the rate of return was too low in terms of capital cost. This, however, ignored the command economy of the USSR and most notably the extra value to Russia of the hard currency to be gained from gas sales.

Yet the sheer size of the pipeline reveals a significant problem within the international gas industry generally, which still causes problems today. Two years after the initial Soviet deal, the Germans signed an agreement with Philips for 5 Bcm/y of gas from the Norwegian sector of the North Sea.

Combining these two deals alone, by 1974 Germany was importing as much natural gas as its total combined coal gas and natural gas consumption nine years earlier. If natural gas was not a key fuel of the future, Ruhrgas and its fellow utilities were, in effect, taking a very big risk. To cover the expense of the considerable pipeline infrastructure needed to supply such imports, initial agreements had to be large. This in turn required an expanded market to absorb the rapidly increasing supply. This 'chicken and egg' problem plays a major role in pipeline gas development. For example, there has been discussion for years about bringing natural gas reserves from Alaska and the northern Canadian arctic to the market in the lower 48 states of the US. However the volumes required to make the investment worthwhile amount to over 10 Bcm/y. This, if delivered would, in the initial period at least, rapidly lower the price of gas in the target market thus reducing the incentive to build the line in the first place.

It is thus not really a surprise that Western Europe's gas industry became dominated at the time by large companies and corporations: British Gas, Gasunie, Gaz de France, Ruhrgas and SNAM. This later became a major question mark over the European Union's drive for greater competition in energy markets, led by the British. Nonetheless in the early years the need was for big deals and long-term contracts. Yet at the beginning of the 1980s, so rapid was the expansion of the natural gas market in Western Europe that the utilities concerned combined to maximize their buying power. As the Norwegian sector began to come on stream, Ruhrgas had combined with Gasunie to buy 5 Bcm/y of gas from the Ekofisk Field in 1975. In 1982, it further combined with its German rival Thyssen with Gasunie and Gaz de France to partly fund an 820 mile under-sea-pipeline to the Statfjord Field in the Norwegian sector of the North Sea. By 1986, with the Troll Field agreement, the deal was that this consortium would import a minimum of 13.5 Bcm/y. With German households switching from oil and coal for heating to gas at the rate of 300,000 a year, Ruhrgas for one could see an ever-expanding market.

The market was expanding elsewhere in Western Europe. If the Germans looked east and north, the Italians and Spanish looked south. In 1977 Italy's ENI signed a deal with Algeria's Sonatrach for the supply of gas with volumes set to rise to 12.4 Bcm/y by 1984, when a TransMed pipeline was to be completed. This went via Tunisia, under the Mediterranean, rising first on Sicily to surface at the southern toe of the Italian mainland. By contrast, with a much less developed natural gas market, Spain opted, until the Medgaz pipeline was completed in 2008, for LNG imports from Algeria (Hayes 2004).

Drivers behind the expansion

Before exploring the creation of the international liquid natural gas industry in the late 1960s and 1970s and its subsequent expansion, it is worth examining why the Western European gas market expanded in the way it did in the 1970s and 1980s. The first and obvious factor was the massive increases in oil prices in the 1970s in two major bursts, 1973 and 1979. While the 1973 increase was not overtly political, being the consequence initially of President Nixon's abandonment of the permit and quota system for all oil imports into the US market, it did become so, with the Arab-Israeli war and the threat of OPEC boycotts. While natural gas prices were often pegged to oil product prices, diversification away from oil dependency was seen as a major requirement for security of supply. Not only did energy prices rise, but the wisdom of using so much heavy fuel oil to generate electricity came under scrutiny. It produced a revolution in electricity-generating technology.

The result was the combined-cycle gas turbine power station (CCGT). Aero-derivative single gas turbines had been used for 'peak shaving' electricity demand for some time. (Power engineers hate the term 'aero-derivative' not least because jet aero-engines are designed to provide thrust for movement and movement is the enemy of power-turbines.) What was different about the CCGT was that it harnessed the waste heat produced by two gas turbines and used it to raise steam for an additional traditional steam turbine. The result was a major increase in the thermal efficiency of power stations. All things being equal, whereas a traditional single-cycle steam turbine power station had a thermal efficiency of around 33 per cent, the CCGT increased this to over 50 per cent and sometimes up to 60 per cent. Very rapidly, gas became the fuel of choice in the power sector, not least because it had other advantages as well as higher efficiency. At a time when 'acid rain' from coal and heavy oil burning was becoming a significant environmental issue, natural gas as a power generation fuel not only reduced sulphur dioxide but produced less carbon dioxide as well.

And it was also far more flexible. In contrast with the coal or oil-fired units, which required days to start up, or nuclear capacity, which was dangerous to run at low to medium power, CCGTs could almost 'follow load'. They took little time to start up and, depending on the range of units, could build up to full load and slowly reduce it (Keay 2006). Very rapidly, across Europe and US, CCGT technology was introduced into the power sector, limited only by the limited number of turbine makers available. While they could also run on LPG and jet kerosene, the natural fuel of choice became pipeline-distributed gas. CCGT gave Japan and other Asian countries a major incentive to seek for natural gas supplies and that meant LNG (Watson 1997).

The LNG revolution

The story of the development of LNG is a heroic one, encountering prejudice, considerable technical difficulties and, not least high expenditure in a highly fluctuating market. When cooled to $-162°C$, the gases in natural gas – mostly methane – liquefy and as a liquid occupy 1/600th of the volume of the gases at atmospheric pressure, thus increasing their energy density enormously and making them much easier to transport in bulk. At least part of the prejudice was historical. Methane had first been liquefied on any scale in 1917, but was first used for 'peak-shaving' in the US gas market in 1941 at a plant in Cleveland, Ohio. Four years after its inauguration, the 3.5 per cent nickel-steel in a storage tank failed and the liquefied gas flowed downwards into a sewer. The resulting fire and explosion killed 128 people, injured over 200 and devastated around a square mile of the city, much of which was flattened. The gas ignited around the leaking tank and the subsequent fire destroyed a second tank. One sewer manhole cover was found over a mile away having been blown sky-high. Given this kind of unhappy publicity, it was hardly surprising that shipping such a material on the high seas did not seem such a great idea.

However, as the US Bureau of Mines investigation pointed out, lessons from the 1944 Cleveland LNG disaster were not as straightforward as the simple idea that LNG was a bomb material about to detonate. The crucial issue here was that only the gas in the sewers actually detonated. The rest of it, around the tanks, merely caught fire. Equally, it was not entirely sensible to keep two tanks close together – without bunds – next door to a railroad marshalling yard and in a residential area with a sewer network. Finally if 3.5 per cent nickel-steel cracked at $-162°C$, 9 per cent nickel-steel did not.

In fact, LNG is remarkably difficult to detonate. The detonation in Cleveland was the result of the mixture of gases and air in the sewers, effectively creating the necessary mixture. Equally, it is not easy to ignite since the flammability range of methane is 5–15 per cent. In effect, if the amount of methane in the air is less than 5 per cent it is too lean to burn and if above 15 per cent it is too rich to do so. Consequently, the liquid gas must vaporize and disperse to these levels before it can even burn. In the case of LNG, this involves a considerable increase in temperature. As a general rule, as numerous experiments first on Lake Charles, Louisiana in 1960 and later on Maplin Sands, Essex suggested, released LNG would burn slowly around the edges.

This does not mean that it is not hazardous. Being so cold, it will crack ordinary steel and will asphyxiate anybody caught within its cloud. Equally as it vaporizes it can expand 600-fold extremely rapidly under certain conditions; a process known as 'rapid phase transition'. This causes a blast wave rather than an explosion. In this regard, LNG is much less dangerous than LPG, for not only is LPG under pressure and thus liable to vaporize much more quickly, but it is heavier than air, so that it 'pancakes' on the floor, creating the exact conditions for a instant detonation known as a boiling liquid expanding vapour explosion (BLEVE). LPG flash fires and explosions from cooking are responsible for a huge number of burns in children in the developing world and LPG is regarded as one of the most dangerous materials in any refinery tank farm.

Even so, in January 1959, the first seaborne cargo of LNG across the Atlantic from Lake Charles to the UK's Canvey Island in the *Methane Pioneer* was told to keep out of all shipping lanes and consequently took 27 days to arrive. However, the team, led by the brilliant British Engineer Denis Rooke, proved that shipment by sea was viable, even if the converted former wartime liberty ship carried only 5000 cubic metres of liquid and was insulated with plywood and urethane. The British Gas Council followed up this with a deal to start importing LNG from a newly independent Algeria to Canvey Island in 1964, shortly followed by the French, who had helped build the world's first commercial scale LNG liquefaction unit – GL4Z – at Skikkda and a receiving terminal at Le Havre. Four years later, the US started to export LNG to Japan from Kenai in Alaska, with Tokyo Electric and Tokyo Gas sharing a 132 million cubic metres of natural gas per year.

This latter was really the first large commercial LNG deal and started the growth of the Far Eastern markets that still dominate the global LNG trade to this day. There were several drivers in this process. First and foremost, Japan and its neighbours in the Far East had no previous access to natural gas by pipeline, while its power industry was also keen to develop CCGT power generation, a process given a huge incentive by the 1973 oil price increases. Secondly, for the first time, a use could be found for what was defined as 'stranded gas' (Van Groenendaal 1999).

This was a worldwide issue. A significant proportion of the global gas being produced was associated with oil. In effect the oil could not be produced without producing the gas as well. In some cases, this gas was re-injected to sustain oil reservoir pressures, but in most places it was flared for lack of an economically viable market. Even prior to fears about climate change, this was a terrific waste. Indeed the oil industry at the time regarded the discovery of natural gas reserves as a nuisance. Flare stacks in Saudi Arabia were supposedly visible from space. Many OPEC members, notably Nigeria, Indonesia, Qatar, Iran and Algeria itself were in fact stronger in gas reserves than in oil, but had no market.

By the early 1980s Japan's imports of LNG were growing at the rate of 5–10 per cent a year, aided by two large liquefaction new-production trains coming on line in 1977 in Indonesia's Kalimantan and Aceh provinces; places with zero local gas markets. By 1984, 72 per cent of global production was going into Japan, of which three-quarters was used for power generation (Stern 2008). This came at a price, however. A liquefaction plant could cost between $1 and 3 billion depending on size, while a single ship could cost $250 million in money of the day. At least part of this high cost related to the handling of such a cold material. The ships, for example, needed very thin membrane tanks well protected by insulation that could contract and expand as the LNG at $-162°C$ flows in and out and the tanks cool and then warm up. Equally, the cooling process that liquefies the gas in the first place is very energy intensive.

Consequently, sellers needed the assurance of long-term sales at oil-related prices with take-or-pay deals and Japan and South Korea were willing to comply with 20 years contracts. The trade used dedicated ships on long-term charters on regular routes. As a result, LNG shipping has a remarkably good safety record. Since 1959 there have been no fatalities on board LNG vessels and very few accidents. In the last 40-odd years, there have been more than 40,000 voyages, involving a fleet now including 337 vessels. There have been a few groundings and collisions, the most bizarre of which was when the USS Oklahoma nuclear-powered submarine surfaced near Gibraltar underneath the LNG Carrier Norman Lady near Gibraltar. Its periscope punctured the outer hull. However, these relatively minor accidents have reinforced the safety record because very little LNG has actually been spilt.

Yet if things went comparatively smoothly in the Far East, matters were not so assured in the Atlantic basin. The US, with it highly volatile gas market, started the LNG era with considerable enthusiasm in spite of a much greater concern over safety than elsewhere. The late 1970s saw a period where the country was widely expected to be facing a significant shortage of natural gas and LNG contracts with Algeria were put in place. The first LNG terminal at Everette, near Boston, started importing in 1971 and by 1978, two other terminals – at Cove Point, Maryland and Elba Island, Georgia – had opened.

However the US market was much more diverse than that of Japan, highly competitive and much less willing to agree to the long-term contracts. At the turn of the 1980s, not only did reserve additions start to equal and exceed production, but the Canadians reduced prices. This 'gas bubble' more or less destroyed the US LNG import business, with a brand new terminal at Lake Charles, Louisiana opening in 1981 to a fanfare and being mothballed the following year. This in turn had serious consequences for Algeria in particular, which had relied on the US market and thus ended up with a significant surplus of expensive liquefaction capacity. A number of ship owners also had their fingers burnt. This pattern of US attitudes to LNG has since happened again. A widespread concern over potential shortages in the late 1990s again saw a spurt of LNG activity, with new terminal projects and – elsewhere – the development of new sources of supply, as in the case of Trinidad. This enthusiasm has subsequently been dashed by the discovery of large reserves of domestic 'unconventional gas' in the first decade of the new century.

Nonetheless, by the beginning of 2011, there were 30 liquefaction plants in operation in 18 countries, with a further six under construction. These send supplies to 79 regassification terminals worldwide. The longest current route is that from the Arabian Gulf to the US Gulf, but this is rarely used. The longest regular route is from Qatar and Abu Dhabi to Tokyo, where an LNG tanker is now unloaded every 20 hours, to keep Japan's power sector going. Ships have graduated from the 5000 cubic metres of the *Methane Pioneer* to the 266,000 cubic metres Qatar-Maximum *Mozah* launched in 2008.

In 2009, some 243 billion cubic metres of gas were delivered by ship or about 27.7 per cent of all internationally traded gas. Most predictions suggest that this will continue to rise, not least because both China and India have now become significant importers. Exporters now include Australia, Malaysia, Thailand as well as those that cluster around the Arabian Gulf.

The contrasting experiences between the Far Eastern markets of Japan, South Korea and Taiwan, which started importing in 1990 and those of US and to a lesser extent Europe reflect a debate within the LNG business. The early pioneers clearly saw the necessity of long-term inflexible price contracts to build up the industry, but did look at the potential that it might ultimately emerge like the oil tanker market, with significant volumes of spot cargoes shipped on single-voyage charters. In effect, as in the oil industry, LNG could be sold while it was at sea and sent to an alternative terminal, thus making its supply much more flexible and potentially more competitive.

In practice, it is very difficult to chart this shift towards an LNG spot market, the only real indicator being the number of single voyage charters for the ships. Estimates now suggest that some 20 per cent of LNG is spot traded, largely outside Asia, although it is thought to fluctuate widely. The most notable shift into spot cargo purchasing occurred when one of Tokyo Electric Power's 8000 MW nuclear power stations was hit by an earthquake in July 2007. This necessitated increased LNG supplies for power generation. Significantly however, the industry had little difficulty in meeting the challenge. Certainly the LNG business is moving towards greater flexibility. The decision to develop a liquefaction plant in Trinidad and Tobago was heavily influenced by its position, ready to supply both Western Europe and the southern US.

By 2010, the four main importers of LNG used 14 suppliers, with five additional sources added in the past five years. The actual supplies amounted to 55 Bcm in a 22.6 per cent growth over 2009. Qatar is now the largest supplier, exporting to 19 countries. As a result, the rates of expansion in LNG capacity is four times that of pipeline capacity and its share of international exports has grown to 31 per cent compared to 23 per cent in 2005. Such is the increased level of flexibility that the US has begun to re-export LNG as surplus to requirements.

The increasing dominance of some Middle Eastern countries, notably Qatar and Abu Dhabi raises the question of whether with its significantly lower price to oil makes it sensible to export LNG at all. Certainly some nearby countries, notably Dubai, might well provide closer markets that could be supplied by pipeline. The answer is probably similar to the motivations of the Soviet Union in the decision to export through the Siberian lines, namely the accumulation of foreign hard currency. This may change, for just as many members of OPEC have discovered that internal demand limits the amount of oil available for export, many countries like Egypt may see domestic requirements increase to the point where exports diminish.

Expansion of unconventional gas

The idea that the US was once again wide open for LNG imports looked like a real winner in the late 1990s. Local reserves in 1998 fell to their lowest at around 4.5 trillion cubic metres (Tcm), with both onshore and offshore reserves in decline. Prices, although highly seasonal were beginning to rise, hitting a high of an average of $15 per 1000 cubic feet in 2005 compared with $2 in the mid-1990s. A further 13 LNG receiving terminals were planned.

The surprise that put all this into reverse was the expansion of 'unconventional gas'. Essentially this was gas from coal mines, tight gas pockets and, most importantly shale rock, previously regarded as too tightly compacted and impermeable to release the reserves that were known to be there. For some decades, some petroleum geologists led by Professor Steve Holditch, of Texas AM University had believed that petroleum resources were 'log normal'. In essence this meant that although gas and oil from conventional reservoirs might be running out, there was much more hidden in rock that was regarded as too difficult to produce. In short, the graph of the potentially available resource followed an exponential curve upwards, with the caveat that as it went up, the resource got progressively more difficult to extract. The technology involved was hydraulic fracturing or 'fracking', which had been around for some time, primarily as a mechanism to release oil from elderly and declining wells. Fracking basically involved fracturing the shale rock with high-pressure water to allow the gas, tightly held in small pockets to flow upward, with the pathways held open with sand and other chemicals.

In the US natural gas business, Holditch, after a long period of being disbelieved, was proved spectacularly correct. By 2008, the Federal Energy Regulatory Commission was putting economically recoverable gas reserves in the US at around 21 Tcm, or roughly four times conventional reserves. Even states like Pennsylvania, widely regarded as played out with regard to hydrocarbons, was suddenly proclaiming new resources. Globally by 2007, the US National Petroleum Council was putting global resources of unconventional gas as high as 914 Tcm, compared with conventional reserves estimated at around 180 Tcm (Yergin 2011).

In Europe during 2009 the CEO of ExxonMobil, Rex Tillerson, stunned the gas business by announcing on the fiftieth anniversary of production from the Dutch Groningen field that the field would continue producing at the same volumes for the next 50 years. As the Dutch government had been concerned that its gas resource was running out, this was a real surprise. The field, it turned out, was sitting above a seam of coal-bed methane and underneath both a shale seam and a stratum full of tight gas pockets. It was not going to run out any time soon. Elsewhere, Poland in particular is seen as having huge potential resources, in spite of the fact that it is currently largely dependent on Russian sources for its existing supplies. The arrival of unconventional gas resources thus has implications for concerns about energy security.

However 'fracking' is not without controversy. In the first place, it uses significant quantities of water to force open the rock, while that water has within it quantities of chemicals and sand to keep open the fractures that have been created. Equally, this water, unless re-used, has to be treated for impurities before it can be released, or it will contaminate rivers and other water. For not only will it contain the original chemicals, but it will also bring with it trace quantities of hydrocarbon, heavy metals, benzene and various other carcinogens. A related problem is that unknown to the driller it may infiltrate the water table and local aquifers used for human consumption. A number of US legal cases are now pending relating to contaminated tap water. In addition there is a fear that 'fracking' may follow the law of diminishing returns, that is more and more water and chemicals may be required to keep up the initial flow. In short, it may in itself be 'log normal' and the preliminary measurement of reserves misleading.

Liberalizing the European gas market

As noted, the expense and long-term contracts required in developing the international links created large gas corporations within Europe, like British Gas, Gaz de France, Gasunie, Snam and Ruhrgas. This ran contrary to the ideals of the European Union whose founding treaty in 1957 had envisaged a single but competitive market in electricity and gas. The aim was for gas-to-gas competition right across Europe. Reinforced by both the treaties of Europe (1985) and Maastricht (1992), this drive finally surfaced in European Gas Directive of 1998.The idea was pushed by the British who had lucratively privatized the British Gas Corporation as a semi-monopoly in 1986 and then split it up into Centrica and BG plc in 1997. The Gas Directive required the 'unbundling' of the integrated companies or shifting the ownership of the gas infrastructure from suppliers. It demanded 'third party access (TPA)' to pipelines. It also required the creation of regulators to grant licenses and insure the efficient functioning of the market. The initial opening of competition was to allow big customers to choose their gas supplier so that at least 20 per cent of each nation's market would be liberalized. This was to be raised to a minimum of 28 per cent after five years and 33 per cent five years after that.

This was not as popular in the rest of Europe as it appeared to be with the British government. Some countries had less pipeline infrastructure and the directive threatened existing take-or-pay contracts and public service obligations. Consequently it was not implemented until 2000 with many negotiated exemptions. In addition, the plan was obviously much more difficult for the former Soviet Union satellites to apply and these were now queuing up to join the EU. A second directive was produced to seek for full competition in the industrial market by 2004 and the total market by 2005. 'Unbundling' was to become full legal separation, while major transit lines were also to be subject to full TPA, except of course if they were brand new (Rutledge and Wright 2011).

In practice full gas-to-gas competition was extraordinarily illusive. First, although the British among others had taken the process much further than legally required, a detailed study by the commission itself in 2004 suggested that the consumer had not benefited much at all (DG Competition Report 2007). Indeed gas prices had more to do with its pricing relationship with oil than competition. Customers did not switch suppliers and small importers were inhibited. Worse still, TPA was inhibiting new infrastructure from being built, the industry was still concentrated and prices were not falling.

Predictably the commission espoused more of the same ideologically defined medicine and decided that long-term 'take-or-pay' contracts were the problem. Then along came the abrupt curtailment of Russian supplies to Ukraine in the winter of 2008–9. While this was in fact the product of a local dispute and not geopolitics, gas supply security suddenly became the talk of politicians. Consequently diversification of supply abruptly appeared more important than residential competition. Equally, such an increase in diversity of supply demanded considerably increased infrastructure expenditure on storage, LNG terminals and pipelines; the very same investments that were being inhibited by TPA. In practice, the Ukrainian threat of embargo is a very rare example of such an action in the global trade in gas. If anything, such disruptions go against the interests of both supplier and customer. While the customer might appear to be over a barrel in terms of the economic consequences, in practice, the supplier suffers not merely from a loss of revenue but also from a reputation for unreliability. The impact of the threatened Ukrainian was – as noted – to increase the diversity of supply as demanded by European politicians (Yafimara 2011).

The commission still seems to believe that liberalising the market is still on track and will eventually emerge across Europe fragmenting the larger utilities as competition blooms. It also appears to be believed that external gas suppliers will adapt their own contractual sales systems to oblige the large numbers of small penny-packet importers, which should result. Given the structure of say, Gazprom in Russia or Sonatrach in Algeria, there is no evidence of this whatsoever. However, that said, the Europeans have been lucky with the boom in LNG supplies, which has been displacing pipeline gas from Russia, with Europe taking a record volume in 2010, around 27 per cent more than in 2009. To a great extent this defuses the potential fear of any price manipulation by the Russians. Nonetheless with declining North Sea production and a growth in demand of 7.7 per cent in 2010, the issue of new supplies will remain on the political agenda.

One aspect of this lies in the continuing role of natural gas in the power sector. The environmental agenda is driving the electricity industry towards the greater use of renewable energy, particularly wind. As has been widely noted, wind energy tends to be highly fluctuating, requiring considerable back-up capacity and this is likely to be gas fuelled. Ireland, with its high target of 33–45 per cent of electricity production from renewables by 2020, has not only been investing in wind farms but has also been replacing old coal and oil-fired capacity with gas-fired power plants, largely because of the fluctuating supply created by wind. In its first annual report on renewables, EirGrid proudly pointed to a graph in which – on Monday, 5 April 2010 – 'wind penetration' of the system for the whole island of Ireland reached a record 42 per cent at around 4.30 a.m. (Eirgrid 2010). However some 16 hours earlier showed 'wind penetration' at around 2.5 per cent at midday. Only gas-fired capacity can cope with fluctuations of this kind. If Europe continues on the path towards smarter grids and renewables and with nuclear out of flavour, the preferred choice of power generation is likely to remain natural gas.

Conclusions

Natural gas provides around 24 per cent of global primary energy by comparison with oil's share of 33 per cent. It is, however, growing faster and its reserves to production ratio at 58.6 years compares with 46.2 years for oil, but is probably underestimated, as the standard calculation does not include the unconventional resources now known to be available in the US. In the past 20 years, the grand total of conventional reserves has grown from 125.7 trillion cubic metres to 187.1 trillion. While the Middle East and the Former Soviet Union have around 72 per cent of these reserves, the invention of LNG and the greatly improved pipeline technology have made such resources accessible to the wider world. With the exception of concerns over Russian supplies with Ukraine and the loss of Libyan supplies during 2010, the gas trade, unlike the oil business, has been remarkably free of political problems. Over the years, the industry has moved from its local foundations to be a global industry. It has moved through a phase of state involvement, which was probably needed to make the switch from coal gas to natural gas but is now, at the consumer end at least, privately owned.

In terms of geopolitics, the natural gas trade has, with the exception of Soviet supplies in the 1970s and Russian supplies to Ukraine in 2008, been remarkably free of the controversies suffered by the oil trade. Even during the terrible conflict in Algeria, its role as a supplier to Europe remained in place. Indeed, while terrorists have often threatened to attack natural gas infrastructure, they rarely actually do so. Equally, while gas utilities do attract consumer anger over increasing prices, this is largely connected with the general inflation in energy prices, notably electricity. It does not spark the outrage that often afflicts the oil industry from motorists in terms of gasoline and diesel prices. This may change in the future, particularly in relation to electricity prices. Here the world has a significant dilemma. While the need to cut carbon output has increased the contribution of renewable energy like wind, it remains highly fluctuating in its supply. Alternative forms of electricity generation, like nuclear, are insufficiently flexible to fill the gaps in output provided by wind and neither can oil or coal. Only gas turbines can rapidly respond as the supply from renewables goes down. As a result, gas is likely to increase its role in power production as the century unfolds, thus increasing demand and prices.

So far the rapid rise in consumption of the fuel has largely gone unnoticed. The fact that gas provides a quarter of global primary energy and is a major international industry remains largely unexamined by academics and journalists alike. This too is likely to change, for natural gas is likely to increase its role in fuelling the global economy, not least because reserves are now known to be far more extensive than crude oil.

Part II
Theoretical Frameworks

4
Geopolitics and International Relations of Resources

Roland Dannreuther

Geopolitics and international relations have always been integral to the drive to secure access to vital global resources. The histories of oil, gas and minerals, as set out in the previous chapters, provide many examples of how the struggle to secure these resources has both excited and driven international politics. The European imperial expansion, such as the late-nineteenth century 'scramble for Africa' was closely linked to the need to obtain the raw materials required for the industrialization and great power ambitions of the European states. The two world wars had important resource dimensions, such as gaining control over the iron reserves in Lorraine and the oil fields in the Caucasus. The Cold War had its resource-linked conflicts over access to the oil reserves in the Persian Gulf and to key strategic minerals, such as chromium, whose reserves were located in the Soviet Union and South Africa. The post-Second World War process of decolonization was frequently linked to international conflict over control of the natural resources found in these countries. From the 1970s onwards, the concept of 'energy security' has been increasingly used in defining the international politics of energy and ensuring that energy concerns have been increasingly seen as an integral part of 'high politics'.

Geopolitics and international relations are intimately linked to energy and minerals security for three principal reasons. The first is the physical fact that fossil fuels and minerals are distributed unevenly around the world and, due to this unequal distribution, some countries and regions are markedly more resource-rich than others. The second is that these resources are critically important for global economic development and growth and are valuable internationally traded commodities. The high levels of rent that possession of these resources can provide, particularly in relation to oil, contributes significantly to international competition. The third factor is the anarchic nature of the international political realm where, due to the absence of an authoritative global government, the relative power and behaviour of states often tend to determine political and economic outcomes more than regional and international governance institutions and international law.

It is the central concern of the discipline of International Relations (IR) to analyse and to seek to understand the anarchic realm of international politics. As a discipline, IR is distinctive in its ambition to construct general theories that identify the key regularities and patterns of interaction in the realm of international politics. These theories have, whether explicitly or implicitly, a normative dimension – they say something about how international politics should be conducted and what the world should look like. They have also tended to present themselves as competing and incommensurate, as representing alternative 'paradigms', and their historical evolution often involves critiquing the perceived flaws of alternative theories. During the Cold War, IR theories tended to fall into three broad schools – realism, liberalism and radicalism or Marxism. Since the end of the Cold War, the theoretical terrain has become more diffuse and complicated with the rise of social constructivism as a critique of both realism and liberalism and with the development of multiple 'critical' theoretical approaches, such as feminism, historical sociology and post-structuralism.

This chapter will assess the ways in which IR as a discipline, particularly its theoretical frameworks, provide insights for thinking about and conceptualizing the geopolitics and international relations of natural resources. The chapter follows the traditional IR theoretical division between realism, liberalism and radicalism and how these provide differing perspectives on the global politics of oil, gas and minerals. This use of the traditional tripartite division of IR theories has the advantage of conceptual simplicity but inevitably does obscure some of the more complex and nuanced theoretical developments in the discipline. But where alternative theoretical approaches do provide additional insights, these will be incorporated into the analysis.

It should be also noted that the direct application of IR theories to the understanding of international politics of energy and minerals has been surprisingly limited. As Brenda Shaffer (2009: 18) points out, the principal journal in IR security studies, *International Security*, has only published eight articles devoted to energy in its 30-year history. However, policy-related IR journals such as *Foreign Affairs*, *Foreign Policy*, *International Affairs* and *Washington Quarterly* have regularly included articles concerning energy and mineral politics. These and other IR contributions, though generally eschewing direct theorizing, are implicitly theoretical, with the main arguments and policy prescriptions underpinned by certain fundamental theoretical assumptions.

The argument of the chapter is that the principal IR theoretical traditions provide differing but also complementary insights and perspectives on the international politics of energy and minerals. As a composite whole they provide the main critical perspectives that academics, analysts and policymakers generally utilize to frame the international politics of oil, gas and minerals. The chapter concludes by assessing the principal contemporary concerns and issues concerning actual and potential conflicts over access to the vital natural resources. Following the main theme of this book, the issue of 'state capitalism' and the implied increased role of the state in energy and mineral markets will be addressed from an international relations perspective.

Realism and geopolitics

For the realist tradition in IR, the intervention of states in resource politics for security reasons is not surprising. The realist tradition is distinctive for the importance that it accords to the state as the key actor in international politics and for the primacy of security in the strategic priorities of states. For realists, it is quite natural that states should be deeply involved in ensuring the security of supplies of vital strategic resources. This is driven by the central insight of realism that there is a fundamental disjuncture between the politics within states and the politics between states, with the latter lacking the overarching sovereign arbiter which is able authoritatively to repress the inexorable drive for power and natural human tendency for aggression. The logical consequence of this condition of anarchy is that the international realm is characterized by distrust and the ever-present prospect of armed conflict and war. Classical realism developed in the first half of the twentieth century as a critique of inter-war liberalism (or idealism) and as a reaction to the optimism expressed by many liberals at that time that international politics could be transformed through international law and the creation of international institutions such as the League of Nations (for key classical realist texts, see Carr 1946; Morgenthau 1960; Niebuhr 1960). In 1979, Kenneth Waltz produced a seminal book which developed realism into a more rigorous social scientific form, known as neo-realism. This established a parsimonious general theory of the international politics as an anarchical system, as structurally determined by the distribution of power between states, and where the internal nature of the state, whether it is democratic or authoritarian, is argued to have no material impact on international behaviour (Waltz 1979).

The failure of international liberalism in the inter-war period and the subsequent development of the Cold War as a systemic militarized and ideological conflict meant that realism has tended to be the dominant theory in IR. Waltz's refinement of neo-realism sustained the intellectual development of realism among IR scholars. But realism can also be thought as a broader tradition, drawing from the older tradition of *realpolitik*, associated with Machiavelli, which highlights the need for statesmen and politicians to promote the interests of state and to place strategic calculation over moral obligation if survival and success are to be ensured. The European state system during most of the eighteenth and nineteenth century was driven by the overarching necessity to preserve the 'balance of power' which is a quintessentially realist concept. The twentieth-century realist scholar and statesman Henry Kissinger explicitly sought to draw from the European experience of the balance of power to develop US foreign policy during the Cold War and during his period on office in the late 1960s and 1970s (Kissinger 1979, 1982). Realism also draws from the influential tradition of geopolitics that developed with the discipline of political geography from the late nineteenth century. These include the seminal works of people like Alfred Thayer Mahan (1890), Halford Mackinder (1919), Karl Haushofer (2002) and more recently Harold and Margaret Sprout (1971) and Ronnie Lipschutz (1989). This tradition highlights the spatial dimensions of state power and inter-state competition and the struggle for influence and control of critical geographical and geopolitical spaces, whether that be the Eurasian heartland favoured by Mackinder or control of the ocean spaces as promoted by Mahan.

The power and attraction of this broad realist tradition is evident in the approach taken by many politicians, leaders, analysts and commentators to the international politics of energy and minerals. A tendency towards a realist and geopolitical approach is closely tied to the fact that these resources are vital for economic development, are located in some of the most unstable parts of the world, and excite the more self-interested nationalist instincts of states and their leaders. In particular, there are two key assumptions which underpin the dominant geopolitical and realist approach. The first is that access to and control of natural resources, of which fossil fuels and a number of minerals are the most critical, is a key ingredient of national power and thus of national interest. As such, dependence on imports of these resources is generally considered as a significant strategic vulnerability. The second is that these resources are commonly perceived to be scarce and becoming more scarce, as with the popular claims that the supply of oil, gas and minerals are close to or reaching a 'peak' level of production (for a critical review of such peak arguments, see Chapters 9–11). These two assumptions are then taken to lead to states increasingly competing for access and control over these key resources and that armed conflict and war over these resources is becoming more likely, if not inevitable.

Such a geopolitical approach to ensuring supplies of energy and minerals clearly characterized much of the international politics of the Cold War. The US and the Soviet Union competed for allies in the Middle East and, for the US, the potential Soviet threat to Persian Gulf oil supplies was viewed (and formally confirmed in 1980 with the 'Carter Doctrine') to be a critical national security concern. The US alliance with Saudi Arabia was cemented primarily on such energy-security grounds. The US has rarely wavered in ensuring its full support for the Saudi regime against internal and external threats, despite the kingdom's poor record on democracy and human rights.

For most of the Cold War, energy security regularly trumped the Western promotion of such values.

The realist approach and post-Cold War period

The realist and geopolitical approach has not, though, disappeared with the end of the Cold War. This is evident in the most prominent and prolific commentator of international energy politics, Michael Klare (2001, 2004, 2008). Although his work is not framed in theoretical terms, and is predominantly empirical and policy-driven, the implicit framework is clearly realist. His core overarching argument is that the ideological struggles of the Cold War between democracy and communism have been increasingly replaced by the struggle for access and control of valuable natural resources. He challenges Francis Fukuyama's claim of the 'end of history' and Samuel Huntington's thesis that post-Cold war conflict is based on conflicts between civilizations. In contrast, Klare argues that the key source of conflict is now resources and that 'it is resources, not differences in civilizations and identities, that are at the root of most contemporary conflicts' (2004: xii).

According to Klare, there are a number of reasons for this heightened role that resources play in post-Cold War inter-state competition and conflict. There is the evidence, as he sees it, that there is a strong likelihood that oil production has or is close to reaching its peak (2001: 42). At the same time, there is increased demand coming from the fast-growing countries of Asia, most notably India and China, who are adopting a clearly mercantilist and geopolitical approach to access to resources. There is also, on the supply side, the fact some of the most critical resources, most notably oil and gas but also a number of minerals, are located in regions such as the Persian Gulf, Central Asia and Africa where there are weak, fragile states with multiple internal and external sources of conflict and where political and religious extremism is growing. With their increased resource wealth, these countries tend either to implode into civil strife, as in many sub-Saharan Africa states, or to adopt revisionist, anti-Western and authoritarian policies, such as in the Russia, Iran and Venezuela. Although Klare does not suggest a direct causal link, he nevertheless sees a strong connection between oil and the emergence of al-Qaeda and the threat of religiously inspired international terrorism (2001: 82). It is the combination of all these factors which leads Klare to conclude that 'resource wars will become, in the years ahead, the most distinctive feature of the global security environment' (2001: 213).

Klare's overarching realist-inspired thesis is undoubtedly a powerful and persuasive framework which captures the political imagination of many analysts and policymakers. For example, the dynamic of competition and conflict between China and the US, with the implicit threat of armed confrontation, has a resource dimension which policymakers in both countries have clearly expressed. In 2005, the Chinese oil company, CNOOC, was effectively barred from taking over control of the American oil company,

UNOCAL, due to Congressional fears of the national security implications (Andrews-Speed and Dannreuther 2011: 107). With the growth of the Chinese presence in Africa and Latin America, US deputy secretary of state, Robert Zoellick, accused China of trying to 'lock up' supplies and claimed this had led to a 'cauldron of anxiety' around the world about Chinese geostrategic intentions. At the same time, President Hu Jintao proclaimed the 'Malacca Dilemma' which expresses the fear of the Chinese government of the strategic vulnerability to US military pressure of this narrow waterway so vital for Chinese energy and resource supplies (Lanteigne 2008: 144). The return of an emboldened and increasingly anti-Western Russia, buoyed by its oil and gas resources, has also resurrected fears of a 'new Cold War' with increased fears of European vulnerability to Russian energy resources (Lucas 2008).

Journalists and policy analysts have also been attracted by the realist and geopolitical approach to resource conflict. This includes the burgeoning literature on the new 'great game' in Central Asia which pits Russia, China and the West in a zero-sum game for control of the region's energy resources (Blank 1995; Karasac 2002). This is mirrored by the fears of a renewed 'scramble for Africa' driven by the region's oil, gas and mineral resources and which is taken to be the core factor reinvigorating great power competition in Africa (Morris 2006; Taylor 2006; Frynas and Paulo 2007). A geopolitical frame is also frequently used in those cases where critical resources, such as oil and gas, are not clearly demarcated within the sovereign jurisdiction of one state. The belief in China and other countries that there are large-scale energy reserves in the South China Seas, where China and the other littoral states have competing sovereign claims, has generally been seen to increase inter-state conflict in the region. The fears over interruptions to critical sea lines of communication, such as the Malacca Straits and the Strait of Hormuz, are similarly viewed as core strategic and military interests and as potential causes of international conflict and war. The Iranian threat in early 2012 to close the Strait of Hormuz in response to the imposition of tougher EU sanctions is a clear illustration of this.

A further factor behind the post-Cold War attraction of the realist and geopolitical perspective is the renewed concern over resource-driven wars. This concern is particularly associated with the claim of Paul Collier that most of the post-Cold War wars around the world, which are now more often civil wars than inter-state wars, are primarily driven by control over resources. Collier and his collaborators developed a complex statistical argument to demonstrate that most post-Cold War wars are not, as most commonly assumed, caused by 'grievance' issues such as ideological and identity claims, but rather by 'greed' and the material rewards offered by control of valuable natural resources (Collier 2000, 2008). As with Klare's analysis, a key claim is that the ideological drivers of conflict during the Cold War have now been replaced by the more material and instrumental goals of financial advantage and personal and social group enrichment. This sense that war and conflicts are ultimately over control of resources has also had an impact on Western policymaking. Despite all the efforts of the US and British governments, it proved very difficult to persuade people that the 2003 war against Iraq was not ultimately about oil.

Overall, therefore, the realist approach to understanding international politics continues to exercise a strong appeal to those seeking to explain the geopolitics of energy and mineral resources. It is not just analysts like Michael Klare but a wide range of journalists, political leaders, government officials and members of the general public who view global resource politics in essentially a realist manner. And it is important to make the point that, even if the resulting analysis is deemed to be flawed, the fact that there is a widespread belief that this does describe what is happening in the world creates, at least to some degree, that reality. A leading neo-realist, Stephen Walt (1987), noted threats only exist if states *perceive* them as threats. For example, if both the US and China decide to interpret their respective resource-driven policies as seeking to gain a strategic advantage against each other this will in itself create a mutual sense of threat and vulnerability. This might occur even if their resource-driven policies, when considered individually, are actually benign and non-confrontational. Theoretically, this reflects the social constructivist understanding of security where threats are not objectively given but are socially constructed (for classic statement, see Wendt 1992).

There is one other key insight which a social constructivist approach can provide to help understand the conditions in which a geopolitical framing of resource conflict is likely to emerge. This is the role that particular interest groups play in the construction and identification of threats. An influential recent innovation in IR is securitization theory which was developed by Barry Buzan and co-authors in their book *Security: A New Framework for Analysis* (1998). The insights provided by this theory are in identifying the processes through which threats are constructed in international politics. The key concept of securitization refers in their analysis to the 'move that takes politics beyond the established rules of game and frames the issue as either a special kind of politics or above politics' (1998: 23). As such, an issue only becomes 'securitized' if certain 'securitizing actors' successfully convince a wider audience that the issue can no longer be seen as just an economic and political issue but rather represents an existential threat to national security. In the sphere of energy and mineral politics, this highlights the fact that there is much variance over time as to whether access to energy and mineral resources is considered as a security rather than just a political or economic concern. And it is often certain interest groups, such as the military or oil company executives or those supporting a more nationalist political stance, who tend to promote a securitized and geopolitical approach so to advance their particular interests and ambitions.. For example, there is a clear logic for the Chinese People's Liberation Army, as well as the US and Russian armies, to accentuate the potential for resource-driven conflicts as a justification for increased resources.

The liberal approach

The liberal approach to IR is particularly sensitive to the way that such unrepresentative sectional interests can hijack policy. Historically, the liberal internationalist tradition emerged within IR as an anti-realist critique of the causes for the First World War. The liberal claim was that the tragedy and slaughter of that war was ultimately due to the political and military elites pursuing, without regard for public opinion and civil society, realist policies such as the obsession with the balance of power, military expansion, secret diplomacy, nationalist xenophobia, and the lack of respect for international legal norms and institutions.

More broadly, liberal IR scholars reject a number of the core assumptions of realism and neo-realism which, they argue, contribute to the perception of a world characterized by distrust, conflict and war. Central to this critique is the liberal refusal to attribute to the state the unconditional primacy accorded to it by realists. In part, this is due to the fact that, according to liberal analysis, the state is only one actor among many others within the realm of international politics. There are also a host of other actors, such as companies, non-governmental organizations, regional and international institutions, which influence and have a determining impact on the development of international relations. As such, the world does not conform to the 'billiard ball' model of inter-state interaction promoted by realists but rather to a more diffuse web-like structure where states are in continual engagement with a multiplicity of other, often independently powerful, actors (Keohane and Nye 1977).

For liberals, this pluralist web is not just a description of the true underlying reality but is also something to be desired and actively promoted. The danger of a world of over-powerful states is that they undermine the effective operation of the global economy through geopolitical manoeuvring. The proper role of the liberal state is rather to act as a 'nightwatchman' to preserve an autonomous realm for international markets and to consolidate the rules and institutions which permit these markets to operate as efficiently as possible. In the 1970s and 1980s, neo-liberalism was developed more as a rigorous social scientific liberal theory, which argued that, despite the condition of anarchy, international regimes and institutions, based on liberal principles of transparency and legally binding norms, could lead formerly antagonistic actors to adopt cooperative behaviour and promote positive-sum results (Keohane 1984, 1986; Baldwin 1993). The European Union is taken as the paradigmatic example of how such a supra-national institution could develop to promote economic and political interdependence and, in so doing, make war increasingly inconceivable in Europe (Haas 1958).

The liberal tradition in IR also rejects the realist assumption of a radical disjuncture between the domestic and international politics and that different moral and practical principles apply to these two realms. Fundamental liberal principles, most notably the obligation to respect individual autonomy (human rights) and that political institutions should be developed to respect these rights (democracies), are seen to apply equally to international as well as domestic politics. A key argument of liberal IR is that democracies conduct foreign policy differently from authoritarian regimes and that the evidence for this is found in the research that demonstrates that democracies are exceptional in that they do not fight wars against each other, the so-called democratic peace thesis (Russett 1993). Liberals reject, in this regard, the deterministic and pessimistic realist view of international politics as being a form of 'tragic politics' and argue that progress can be made; universal human rights can be promoted and defended; countries can shift from being authoritarian to democratic states; international politics and the global economy is more effective and efficient if conducted between democracies; regional and international norms, regimes and institutions can be developed which are conducive to increased cooperation; global prosperity is best assured if markets are left open and trade is liberalized; and that anarchy can be overcome and war avoided.

The liberal approach to energy and mineral security: The role of the market

In the energy and mineral realm, the overarching claim of analysts adopting a liberal framework is that the realist and geopolitical approach tends to underestimate and mis-describe the role of international markets for these commodities. As set out by Humphreys in Chapter 2, the markets for non-fossil fuel minerals are global, highly competitive and are dominated by large private companies. Historically, this was admittedly not the case for oil markets which, up until the first oil shock in 1973, were dominated by the internal trading schemes of the major Western oil companies and their concessions in oil-exporting countries and most globally traded oil was bound up in long-term bilateral deals between consumer and producer nations. But this was radically transformed in the late 1970s as the effect of the oil exporters' nationalization of domestic production was to break up the existing vertical integration of industry and significantly to increase the fungibility of crude oil and to make price formation more transparent and predictable. The subsequent creation of oil futures contracts and spot oil markets in New York and London cemented the formation of a 'new oil world depending on short-term rather than long-term contracts' (Goldthau and Witte 2009: 376). Gas markets have, it is true, not liberalized to the same extent and remain tied, particularly in Europe, to long-term bilateral deals characterized by destination clauses and prices indexed to oil. But recent years have also seen, as set out in Chapter 4, the emergence of a potentially global market for gas, primarily through the expanding role of liquefied natural gas (LNG) and the impact of the increasing viability of non-conventional gas production.

For liberals, a key flaw in the geopolitical and realist approach is to overestimate the capacity of states to influence and control the operation of these markets, and the implicit rules and norms of cooperation which underpin them. A key example is the oft-cited threat of oil exporters' unsheathing the 'oil weapon' which was first attempted to be used, even then unsuccessfully, in 1973. It is now generally recognized to be almost impossible to impose an oil embargo of this nature as the global oil market is simply not controlled by the producer states. In less liberalized markets, such as gas, the state has certainly greater potential powers but again, from a more liberal market perspective, there is a tendency to exaggerate the potential for conflict and to overlook the reality of cooperation. An example of this is the perceived threat to the European Union of Russia using its gas supplies as a geopolitical tool. The fact that this threat is only perceived to emanate from Russia and other suppliers for the European regional market, which include gas producers in the politically unstable North Africa and Middle East, is itself an implicit confirmation that the gas market is generally seen to be effective. And, in the Russian case, the fact that gas supplies were never interrupted even during the Soviet period, despite extremely high levels of ideological conflict, indicates that there is a mutual interdependency between Russia and Europe in the energy field which constrains Russia's predilection for geopolitical manipulation.

The role that markets generally play in managing energy and mineral security is also, from a liberal perspective, often not well understood by government elites. For example, the emergence in the late 2000s of fears of the potential restriction of supplies of 'critical minerals', such as rare earth materials, ignores the lessons of the previous period of anxiety over mineral supplies in the late 1970s, over the Soviet and South African control over a number of minerals, being resolved, albeit with some time lag, by the industry finding and producing supplies elsewhere. In fact, the resulting oversupply of these minerals contributed to a subsequent collapse in prices. Another example that liberal economists highlight of this tendency to exaggerate scarcity is the whole issue of 'peak' oil, gas and minerals where, as argued in Chapters 9–11, a geological determinism ignores the ways in which markets, aided by market-driven price signals and technological advances, continually discover new reserves and undermine the gloomy peak prognoses.

Depoliticizing the politics of resources

The logical conclusion from this liberal market perspective is that states should spend considerably less time in politicizing and securitizing access to oil, gas and minerals and expend greater efforts at building trust and cooperation through the expansion of liberal norms, regimes and institutions which strengthen energy and mineral markets. The International Energy Agency (IEA) is seen as an important institution which provides rules and framework of cooperation between the energy-consuming states for ensuring protection against short-term supply restrictions. It is argued, in this regard, that China's geopolitical anxieties would be considerably reduced if it joined this organization and assumed a common front with other oil-importing nations. Similarly, the EU has dedicated considerable efforts to encouraging Russia to sign up to the European Charter Treaty (ECT) and to a clearly defined set of rules for investment, transit and trade in the energy sector, complemented by a dispute settlement mechanism. The fact that both China and Russia have now joined the World Trade Organization (WTO), in 2001 and 2011 respectively, is a positive indication of their commitment, in principle, to the liberal multilateral trading system and the provision of mutual market access without barriers. Regional institutions in Asia, as well as Europe, have also sought to address common interests and concerns over supplies of oil, gas and minerals. Another institution to have considerable constructive potential is the International Energy Forum (IEF) as an institution dedicated to fostering consumer–producer dialogue and enhancing mutual understanding and transparency. Although the IEF currently only engages in informal dialogue, it has the potential for identifying the mutual needs and aspirations of consumers and producers and thus facilitate cooperative and mutually beneficial solutions.

Liberal critics of the geopolitical or realist approach are not, though, blind to the weaknesses and fragility of many of these institutions or the reality that geopolitical considerations, and state intervention, are regular features of energy and mineral markets. Most liberal analysts, even if they generally believe that only liberal market-led solutions bring longer-term security, do not deny the multiple illiberal practices and features in the international energy and mineral markets. For example, there is a general recognition that corruption is pervasive in these industries and that Western and other importing countries are, at the very least, complicit in such illiberal practices (McPherson and MacSearraigh 2007). This contributes to the shady world of secret oil fixers, traders and dealers. The historical record of Western political relations with energy-rich states is also hardly one of liberal good practice. The 'special relations' between France and the oil-rich state of Gabon and its corrupt leader, Omar Bongo, which involved a complex tangle of the French oil company, ELF, the French secret services and parts of the political establishment is only one, if one of the most corrupted, examples (Shaxson 2007).

From the liberal perspective, Western liberal countries and their oil and mineral companies are certainly culpable for these illiberal practices. As such, companies are encouraged to assume greater responsibility and to follow the principles of corporate social responsibility and to embrace initiatives for good practice such as the UN Global Compact (Frynas 2009). States are similarly urged to assume a more transparent and open relationship with key producer countries. But the underlying assumption is that the root causes of continuing illiberalism and lack of transparency of energy and mineral markets is not ultimately found with these Western states and companies. It is rather found in the systemic failures of governance and democratic accountability inside many of the resource-rich states in which Western states and companies are forced to operate. The root cause is thus ultimately found in the illiberal pathologies of the resource-rich states.

This location of the source of the problem of the illiberalism of energy and mineral markets as being internal to the resource-producing states is implicitly confirmed by much recent social science research. The body of research which argues that resource-rich states suffer from a 'resource curse' and that they have developed 'rentier' states, which promotes both poor economic policies and authoritarian forms of government, confirms the general thesis that the root causes of the continuing illiberalism of energy and mineral markets is found ultimately within the domestic politics of these states (for classic statements see Karl 1997; Ross 2012). As such, it is a striking confirmation of the liberal conviction that it is the domestic politics of states which structures and defines international politics.

The next chapter by Ostrowski provides an in-depth assessment of the intellectual underpinnings of the 'resource curse' thesis and the development of the 'rentier state' model. What is important in the context of this chapter is to highlight how the entrenched illiberalism of the resource-rich states is seen to be the fundamental cause of the global market imperfections. The key factors are the excessive power accruing to the elites of resource-rich states, the ways in which resource wealth is treated as a personal source of enrichment, the lack of transparency and endemic corruption, and the repression of civil society through undermining any moves towards political liberalization and democratization. It is these domestic practices within the resource-rich states which lead to companies, which would otherwise operate in an open and liberal manner in a properly regulated market, being implicated in corruption and secret deals. It is for similar reasons that Western importing states will also often forego their normal liberal demands, such as over human rights and democracy, when dealing with energy and mineral-rich countries.

From the liberal perspective, the rot starts from within and spreads out to affect the broader political economy. It is for this reason that liberal policy prescriptions continually focus on the need for good governance to overcome the illiberal practices of 'rent-seeking' and the perversions of the 'rentier state' (Campbell 2006; MMSD 2002). It also helps to explain the focus of Western countries on the need to increase transparency and initiatives like the 'Extractive Industries Transparency Initiative' and the 'Publish What You Pay' campaign. It is a strong liberal conviction that openness of information is what mobilizes civil society against the entrenched privileges of state elites. It is thus a key weapon to be used to undermine the foundations of illiberal economic and political practices of the oil, gas and mineral markets (Karl 2005).

Radical and critical approaches

A common and distinctive feature of Marxist-inspired radical approaches, such as dependency theory and critical IR theory, is in their challenging of this liberal claim that the primary responsibility for the conflicts and tensions in energy and mineral markets lies primarily with the producing states. From a radical perspective, this unjustifiably absolves responsibility from Western states and companies for the relations of coercion and structural injustice which are the underlying causes of international resource-based conflicts. The realist approach is also a target of the radical critique but realism is taken almost to be self-evidently flawed since its explanations of international behaviour assume no potential for radical change and thus explicitly condone the structural injustices of the international status quo. Liberalism is thus a more insidious ideological threat since, like the radical approach, it does offer policies for change and reform, based seemingly on altruistic and benign universal principles. But, for radicals and critical theorists, the liberal framework, and the policy prescriptions that flow from this, only perpetuate rather than challenge the underlying deeply unjust structures of international power and domination.

For classical Marxists, this is because liberalism defends and supports rather than condemns global capitalism. For dependency theorists, economic liberalism is a form of imperialism which consolidates rather than dismantles the domination of the North and the oppression of the South. For critical IR theorists, liberals are condemned for taking a 'problem-solving' approach, to use Robert Cox's (1981) term, which means relying on technical solutions to the resolution of problems, such as the adoption of transparency measures or the promotion of corporate social responsibility, rather than adopting a truly 'critical' approach which asks more profound questions of the moral and political legitimacy of the contemporary international system.

Dependency and injustice

A good illustration of the continuing salience of the radical approach in serious academic analysis of international energy issues can be seen in Ray Hinnebusch's textbook *The International Politics of the Middle East* (2003). Hinnebusch adopts a fairly classical Marxist-inspired dependency approach, arguing that the states of the Middle East continue to 'exhibit many of the classic features of dependency' which involves dependence on a few basic export commodities, the most important of which is oil (2003: 35). This dependency is perpetuated by the consistent failure to process these raw materials into high-value finished goods. Politically, this leads to a symbiotic relationship between the core and the locally dominant economic and political classes which detaches these elites from their local populations and inhibits autonomous economic development. A Western-supported military and security structure suppresses all indigenous challenges to these dependent and unjust structures of core–periphery relations. For Hinnebusch, the modern history of the Middle East is a perpetual struggle between local indigenous forms of resistance, such as the radical Arab nationalist struggles in the 1970s, and Western states and multinationals who have continually sought to repress and eliminate such attempts to forge a regional autonomy. The impact of the end of the Cold War on the Middle East is, for Hinnebusch, far from the benign and positive outcome that liberals proclaim. Rather, the collapse of the Soviet Union removed one of the few sources of external support and resistance to US hegemonic political and economic domination of the region (2003: 204).

Dependency theory, and radical Marxist-inspired theories more generally, has undoubtedly been a powerful ideological source of inspiration for the drive to independence and autonomy in the developing postcolonial world. For the international oil and mineral industries, it provided the underlying intellectual foundation for many of the developments of the 1960s and 1970s, such as the policies of nationalization, the formation of OPEC, the attempts to establish a number of mineral cartels, and the more general ambition to construct a New International Economic Order (NIEO). This mood of the need for radical change was well captured in the Brandt Report of 1980 which recommended a drastic shift of resources from the North to the South (Brandt 1980).

The same nationalist, anti-Western and anti-liberal spirit can also be seen to infuse the revival of resource nationalism from the 2000s onwards. This time, though, the ideological support is not generally in defence of socialist autarchy but of authoritarian state capitalism. In Russia, a narrative has been successfully constructed of the privatization of the oil, gas and minerals sectors during the 1990s as part of general Western plot to weaken the power of the state. The partial renationalization of these sectors during the Putin period was presented as a necessary step to reconstruct and strengthen the state and to transform Russia into an 'energy superpower'. In China, the perception that the West is seeking to constrain and prevent the country's 'peaceful rise' has had a significant impact in supporting a more assertive and mercantilist approach to securing its energy and mineral needs through the international activities of its state oil and mineral companies. In Latin America, there has been a wave of nationalizations in the energy and mineral sectors in countries like Venezuela, Bolivia and El Salvador. Similar dynamics of increased state intervention can be found in Africa, Central Asia and other parts of the world. The 2003 US-led intervention into Iraq also considerably strengthened radical critiques of the international politics of energy as the war was widely perceived to be intimately linked to US ambitions to control Middle East oil supplies (e.g. Harvey 2003).

The critical approach

Apart from its continuing power as a political tool to shift the balance of power from the North to the South, such radical critiques have also contributed to more academic contestations of the dominant liberal conceptualizations of the politics of energy and minerals. Particularly in the fields of economic and political geography, there has emerged a number of 'critical' analysts who adopt a wide variety of theories and methodologies, such as post-structuralism and Foucauldian analysis (Watts 2004a; Barry 2006), action-networked theory (Barry 2006; Mitchell 2009) and global production network analysis (Bridge 2008a). These analyses are methodologically and theoretically more nuanced and sophisticated than the cruder simplifications of traditional dependency theory. But they do have a common concern to critique the general ahistoricism and the lack of attention to the complex structures of power which characterizes much liberal analysis of natural resources, most notably seen in the highly influential liberal thesis of the 'resource curse' and the associated presumed ills of resource dependency.

These critiques do not deny that the resource dependency of many developing resource-rich countries is an important factor behind the tensions and problems in the international politics of energy and minerals. But they argue that the conventional liberal approach tends to be too deterministic and too narrowly focused on the predations of the producing states that critically ignores the role of other actors. An example of this tendency towards a causal determinism is seen in a way that statistical studies, such as those produced by Michael Ross (2001, 2012), demonstrating the causal linkages between oil and authoritarianism, are often interpreted to mean that it is the mere physical possession of oil which distorts economic and political developments and perpetuates authoritarian structures of power. The problem is that this ignores the complex set of historical, social and political relations which are essential for understanding the evolution and development of the social and political structures through which a country utilizes its resource wealth. For example, the political economy of oil in Saudi Arabia is inextricably linked to the security concerns and hegemonic penetration of the US in the region and to the continuing support provided by Washington to the theocratic and patrimonial state in exchange for Saudi support for US energy security interests and the interests of US oil multinationals. Liberal prescriptions of the need to promote 'good governance' fail to take into account these broader historically and internationally determined conditions which constrain progressive political change.

Another effect of the liberal focus on the resource-producing state is that it has tended to marginalize the role of the oil and mineral firms in international politics. A 'critical' antidote to this can be seen in the work of Gavin Bridge (2008a) who provides a detailed analysis of the multiple actors involved in what he calls the 'hydrocarbon commodity chain' which extends not just to extraction and production but also further down the chain to refining, transportation and to consumption. He highlights how there are actually many different types of oil firms, how they are all engaged in complex negotiations and bargaining not only with states but with other firms. Certainly, the most politically contested element of the chain tends to be at the exploration and production end but it is important to recognize that at this stage the terms and conditions are not just dictated by states but are the outcome of a complex process of political and economic bargaining between states and companies. There is certainly an extensive literature on the global oil industry, which Bridge's work builds upon, but it is rare that this literature is fully incorporated into more critical political analysis (for example, Turner 1978; Bindemann 1999; Mommer 2002; Noreng 2002; Marcel 2006).

In the same way that the liberal approach tends to marginalize the role of the firm in its conceptualization of the international politics of energy and minerals, the same is the case for local and subnational communities. One of the emphases of dependency theories – the salience of capitalist 'enclaves' in developing countries – is something which has been taken up by anthropologists and geographers studying the specific localized manifestations of the oil and mineral industries (Ferguson 2008; Yessenova 2007). In many resource-rich African states, there are small territorial enclaves of mineral extraction, protected by private armies and security firms, in conditions of weak, insecure states or a more generalized civil war. The complex social and political realities of these 'enclaves' provide powerful illustrations of the interlinkages between international and national companies, national states, foreign states, subnational regions and local communities. The geographer Michael Watts has looked at this in relation to the oil industry in the Niger Delta and identifies the following actors, agents and processes

as not only IOCs, NOCs and their service companies but also the petrostates, the engineering companies and the financial groups, the shadow economies (theft money laundering, drugs, organised crime), the raft of NGOs (human rights CSR groups, monitoring agencies), the research institutes and lobby groups, the landscape of oil consumption and, not least, the oil communities, the military and paramilitary groups, and the social movements which surround the operation of, and the shape and functioning of, the oil industry narrowly construed.

(Watts 2009; see also Watts 2001, 2004a)

Conclusion: Geopolitics, IR and the future

Overall, the three broad IR theoretical traditions of realism, liberalism and radicalism provide differing but also complementary insights and perspectives on the international politics of energy and minerals. The realist tradition emphasizes the strategic interests that states have in ensuring the security of supplies of such critical resources, the multiple potential sites for contestation and the dangers of conflict and war over access to these resources. The liberal approach highlights the pacifying and conflict-reducing role of markets in the energy and mineral sectors; the extent to which such markets enhance and promote global security and prosperity; and how it is critical to strengthen these markets through regional and international cooperation and by intensifying processes of economic and political liberalization, which is most notably deficient among many resource-producing developing states. The radical and critical tradition highlights the continuing salience of imperial and colonial legacies to the energy and mineral industries, the past and continuing structures of inequality and injustice, and the complex array of actors which are continually engaged in acts of coercion and resistance at the local, national, regional and international levels.

IR as a discipline, and the theories that are developed within the discipline, has also a particular vocation to identify and seek to extrapolate the implications of contemporary developments in international politics. How does IR help us understand the contemporary international situation? Since the mid-2000s one of the key developments which has exercised much thought and anxiety, at least in the West, is the clear evidence of a shift of power towards the emerging countries of China, India, Brazil and other fast-developing and large economies. For much of the post-Cold War period, the dominant view was that the US was the sole remaining superpower and the system was best characterized as unipolar (Wohlforth 1999). Since the mid-2000s, and particularly with the economic crisis in much of the industrialized world since 2008, there is a growing recognition that the system is becoming increasingly multipolar and that countries like China and India will have a much greater impact on international developments (Zakaria 2008). Among IR scholars, there has been a particularly dynamic debate over the rise of China and whether this rise represents a threat to the West, leading potentially to war, or whether it represents an opportunity to ensure the successful integration of China into the liberal international system. There is, though, no consensus about which is the more probable outcome. As Friedberg (2005) notes, realist, liberal and constructivist interpretations are not only divided between themselves but also internally; there are 'optimists' and 'pessimists' over the potential for conflict among realist, liberals and constructivists.

In the more specific realm of the international politics of energy and minerals, there is a related debate about the implications of the growing ambition and purchasing power of the oil and mineral companies from the emerging countries. One distinctive feature of these companies is that they are mainly state-owned and/or have close ties to the home state. The idea that the global capitalism is increasingly characterized by 'state capitalism' is, to a considerable extent, linked to the activities of these companies which are perceived to operate not purely on commercial principles but also have a 'national mission' directed and controlled by their home states (Bremmer 2009; Economist 2012). However, as with the 'China threat' debate, there is no clear consensus on this and sceptics argue whether this represents a fundamental challenge to the liberal capitalist system and that, in reality, these companies are more driven by market than national concerns. But there is nevertheless a growing view that there is some broader ideological shift away from the neo-liberal paradigm of the 1980s and 1990s and that the new paradigm includes a stronger and more interventionist role for the state in the capitalist system. If this is the case, there are significant implications for the global management of the international system which will undoubtedly strain relations between the West and the emerging powers.

Another key concern for IR scholars in seeking to understand contemporary dynamics is the potential for conflict and insecurity to emanate from what Fred Halliday (1981) called the 'arc of crisis' from Central Asia, through the Middle East, to sub-Saharan Africa. This region has remained deeply unstable since decolonization but the post-Cold War emergence of al-Qaeda and Islamist terrorism, the wars in Iraq and Afghanistan, the nuclear ambitions of Iran and the multiple civil wars in Africa have meant that concerns that this broad region represents for international security remain heightened. There is a perceived causal linkage between the rich energy resources of the region, instability and the rise of revisionist and anti-Western sentiments (Friedman 2006). The nuclear proliferation issue in Iran, which in 2012 led to Iranian officials threatening to close the Strait of Hormuz, thereby potentially cutting out one of the main routes for global oil supply, demonstrates that the prospect for conflict and war in the region remains strongly present.

However, there have also been a number of more favourable and positive developments. The overthrow of long-standing dictators in Egypt, Tunisia and Libya during the Arab Spring of 2011 have demonstrated that the oil-rich region of the Middle East is far from immune from the popular pressures for political freedom. In Russia and in Iran, the regimes in power have also faced serious challenges to their popular legitimacy. There is a renewed economic and political confidence in Africa, which is at least partly due to the improvement in the raw commodity sector. The Western interventions into the region during the 2000s, most notably in Iraq and Afghanistan, are being reversed with troops now out of Iraq and planning to be withdrawn from Afghanistan. A degree of stability is returning to Iraq and oil production is gradually returning to its full capacity. But the brutal repression in Syria during early 2012, and the threat of a descent to civil war, suggests that peace and stability are far from assured in the region.

It is also to be recognized that, as this chapter has articulated, the role of the oil, gas and mineral resources of the region play an important role in contributing to both the exacerbation and the resolution of tension and conflict in the region. The causal linkages are not, though, simple. IR theories help to provide a set of broad frameworks and perspectives for analysing and understanding the role that these resources play in the international politics of the region as well as more generally in the world. It is this helpful critical function that the theories of IR fulfil. However, it should be stated that IR is not the only discipline that provides theoretical insights with relevance to the politics of energy and minerals. Political geography, anthropology, sociology, economics and law, to provide a far from exhaustive list, all provide key conceptual and theoretical insights and benefit from their mutual interpenetration and intellectual exchange. Much of the analysis above is actually the result of considerable interdisciplinary interchange and borrowing from different disciplines and approaches. The next chapter examines one of the richest and most influential of such disciplinary approaches – that of the political economy.

5
The Political Economy of Global Resources

Wojciech Ostrowski

The purpose of this chapter is to engage with the issue of conflict and collaboration in relation to global resources from a political economy perspective. It should be noted from the outset that the political economy of resources, an enquiry which primarily looks at the link between a country's development and its minerals and fossil fuels does not constitute a coherent or unified intellectual paradigm. Rather, it is a collection of different debates that have taken place since the middle of the twentieth century in connection to the extractive industries and their socio-political impact. The chapter will argue that each of these debates took place in a specific intellectual context, dealt with different geographical regions of the world and was associated with a diverse range of minerals and fossil fuels. Accordingly, all of those elements have to be properly highlighted if we are to understand why in different historical periods or regimes resources have been associated with either conflict or cooperation.

The study of the political economy of resources originated in the Western Hemisphere in the 1950s and was first associated with the mining industry. Initial studies conducted by North American scholars, within the realm of the staple theory, found a correlation between resources and a country's overall positive development. This encouraging view was challenged by dependency theory, developed by South American academics turned practitioners in the late 1950s and 1960s, who argued that existing patterns of mineral extraction perpetuated structural injustice between the core and peripheral countries and inevitably led to a conflict between the global north and the global south. The determinism of dependency theory was subsequently questioned by development in the postcolonial world in the 1970s with the rise of OPEC and the nationalization of previously foreign-owned extractive industries in parts of North Africa, Middle East and South America by newly assertive states.

The rise of 'third-world states' sparked research by scholars predominantly of the Middle East which began looking at the internal politics of the oil-rich states and the issue of rent. Rentier State Theory pointed to a number of possible conflicts at the national level, which materialized with the onset of the debt crisis in the 1980s and the end of the Cold War. In the heyday of globalization in the 1990s, the focus remained on the domestic politics of the resource-rich countries with one significant difference. The resource curse thesis, which became extremely influential, not only analysed the negative impact of resource wealth on development but also began to seek solutions to address these negative impacts. Sub-Saharan Africa became a key focus of concern and study for those seeking to cure the problems of the 'resource curse'. With the rise of China, India and the 'return' of Russia in the 2000s, the focus yet again shifted from domestic politics back to the international level and the 'great game' for access to resources, primarily oil, played out between the empowered south and the allegedly weakened north.

In order to unpack the different debates that constitute the field of the political economy of resources, this chapter will follow a standard narrative of political economy: classical economy, modernization theory, dependency theory/world-system theory, the rise and crises of the third-world state and the globalization debate. This traditional narrative significantly overlaps with the different historical regimes of oil and minerals as set out in Chapters 1 and 2 respectively. The final section of the chapter will discuss the emergence of 'state capitalism', which has become an important feature of the political economy of resource since the mid-2000s.

Modernization theory and its critique (1940s–60s)

Staple theory

The political economy of resources is a twentieth-century creation. Until the turn of the century, political economists did not directly deal with the issue of minerals, mining and development. This was largely due to the fact that classical economists in the eighteenth and nineteenth centuries were not concerned with problems of underdevelopment and mining was not discussed in connection with national economic growth.

The issue of minerals and development was only indirectly addressed by David Ricardo, an English political economist, in his writings from the beginning of the nineteenth century in which he developed the theory of comparative advantage. The theory stated that every country has an ability to produce or extract a particular good at a lower cost relative to other goods. The exportation of such goods in exchange for other goods results in welfare increases for every country. In the 1930s, Ricardo's theory was redefined by Bertil Ohlin, a Swedish economist and a Nobel Prize winner, who argued that countries should export commodities which they hold in relative abundance. The export of commodities could be a comparative advantage for a country (Eckes 2011: 88–9).

It should be added that throughout the nineteenth century the issue of mining was taken very seriously by those concerned with international trade. According to Jeannette Graulau, this should not come as a surprise since the history of mineral commodity markets from the fifteenth to eighteenth century demonstrated that these were the most profitable commodity trades operating on a vast, we would say today, global scale. The trades had also significant impact in the development of credit instruments in Europe (2008: 137).

While the issue of minerals was largely overlooked by classical economists the comparative advantage paradigm which they crafted was eventually applied to minerals, mining and development in a rigorous fashion. The Canadian staple theory, which in the 1950s and 1960s generated a comprehensive body of research, aimed to explain the significant and largely positive role that natural resources has played in the Canadian economy (Watkins 1963). The framework for analysing the role that natural resources played in the development process in North America was put forward by two Canadian economic historians, Harold Innis (1956) and W. Mackintosh (1964).

The staple theory stated that economies with an abundant supply of accessible natural resources have a meaningful advantage in the development process and that 'rapid progress in new countries is dependent upon the discovery and development of cheap supplies of raw materials' (Mackintosh 1964: 13). The underlying assumption of staple theory was that staples exports are the leading sectors of the economy and set the pace of economic growth. Economic development should be a process of diversification around an export base. Thus, Melville Watkins argued that the central concept of staple theory 'is the spread effects of the export sector, this is, the impact of export activity on domestic economy and society' (1963: 143). The spread effects were divided in four categories:

> [F]orward linkages involving processing of the staple prior to export; back-ward linkages involving the production of inputs such as resource machinery and transportation infrastructure that are required to extract the staple; final-demand linkages involving the production of consumer goods and services to meet the regional needs of those who are employed in the staple industry; and fiscal linkages involving the expenditure of rents and profits generated by resource production.
>
> (Gunton 2003: 68)

Foreign capital and outside technology were crucial to kick-starting these spread effects.

Staple theory, in its ideal form, saw natural resources as something positive to the overall development, progress and cooperation between nations. Staples should permit a country to step firmly on the path of modernization and follow other more developed countries. Thus, in its fundamental assumptions, staple theory complemented modernization theory, which promoted a uniform and evolutionary vision of social, political and economic development and had its heyday in the 1950s and 1960s (Parsons 1951; Rostow 1960). According to modernization theory, all societies, once they begin the process of modernization, must move from development stage A to B, C etc. The elites of the states that go through the process of modernization, guided by the Western model, will adopt its technology, assimilate its values, import its financial, industrial, and most importantly its educational institutions. Modernization theorists also agreed with the 'assumption of economic rationality implicit in the economic growth models of traditional economic theory' (Valenzuela and Valenzuela 1978: 539).

Even so, staple theorists were also aware of the shortcomings of the theory that they were putting forward. Innis argued that regional economies that were overdependent on staples can find themselves in a disadvantaged position vis-à-vis other parts of the country and over time suffer from cyclical crises. In the mid-1950s he argued that in Canada the entire industrial structure and government activities of the staple region had 'become subordinated to the production of the staple for a more highly specialised manufacturing community' (Innis 1956: 385). Thomas Gunton (2003) pointed out that for Innis the process of staple development was characterized by disequilibrium as 'each staple in turn left its stamp, and the shift to new staples invariably produced periods of crisis in which adjustments in the old structures were painfully made and a new pattern created in relation to a new staple' (Innis 1972 quoted in Gunton 2003: 69). Moreover, it was also suggested that in his work Innis indirectly alluded to structural inequalities that developed between the centre and the resources-rich periphery not only at the national but also at the international level (Weaver and Gunton 1982: 8). This links directly to a central theme of dependency theory.

Dependency theory

Two key developments in the overall story of minerals in the first half of the twentieth century were the First World War and the Great Depression. During the First World War the issue of minerals and mining was addressed for the very first time in connection with the needs of the 'war economics' of developed countries. The war also gave birth to the concept of strategic minerals when it became clear in the US that

> domestic supplies of a dozen minerals [including tin, nickel, platinum, nitrates, and potash] were inadequate in quantity or quality or both, another half dozen adequate for peace but insufficient for war, and petroleum production barely sufficient to meet the Nation's normal demand and much too small for the abnormal demands of war.[1]

Thus, the US, a rising power of the day, recognized that some minerals and oil are commodities which have security implications and could easily become an important part of a modern conflict.

The First World War was followed by the Great Depression which further underscored the problematic nature of natural resources, albeit for very different reasons. The collapse of a worldwide liberal order gave birth to a series of studies in the 1940s, which pointed to the inherent unfairness of the classical economic paradigm of comparative advantage. The classical paradigm was first challenged by Raul Prebisch (1950), an Argentinean economist and the father of Latin American structuralism, who in collaboration with other economists from the continent began questioning 'whether the integration of Latin American economies into the internal markets benefited these countries, as the theory of comparative advantages proclaimed' (Graulau 2008: 139). Together with Hans Singer he developed a thesis on the deterioration of the terms of commodity trade (Prebisch-Singer thesis) which showed that countries which followed this path eventually found themselves in a downward spiral (Toye and Toye 2003).

As such, for dependency theorists, the erratic stop and go performance of the Latin American economies was closer to reality than the seamless progression through the ascending stages of development that modernization theory suggested. Prebisch argued that 'underdevelopment' could not and should not be thought of as a temporary historical phase. It was also 'false' to assume that 'the periphery should undergo structural transformations similar to those that had been pursued by the industrial countries' (Mallorquín 2007: 809). Furthermore, the international division of labour did not work to the mutual benefit of all parties as comparative advantage proclaimed. Rather, Celso Monteiro Furtado, a leading Brazilian economist of the twentieth century, argued in the 1950s that 'the classical international division of labour, which confined the periphery [Latin America] to the production of raw materials and the centre to the production of industrial good [primarily the US], perpetuated the process and condition of underdevelopment of the periphery' (Graulau 2008: 141). Thus, it is essential to differentiate between the 'periphery' (primary goods producers) and the 'centre' (industrial nations). Dependency theory, a counter discourse to the theory of comparative advantage, highlighted a conflict between the global south and the global north.

Dependency theory, in striking opposition to modernization theory, argued that the modernization process was not universal and in the final analysis benefited only a powerful minority within a peripheral country which imposed its development patterns and values on the majority. The monopolist multinational enterprises in tandem with local oligarchs pushed for the development of the minerals sector which failed to benefit the larger economy. In other words, the spread effect, as stipulated by the staple theory, did not take place. It was argued that the minerals industry was only interested in investing and developing the infrastructure (roads, ports, bridges, electric plants) which it needed for its operations and which served the external export sector. The service economy that supplied the mineral export sector was a consequence of foreign investment, and never became an integral part of the economic structure. The fact that mining took place in isolated areas, so-called enclaves, added to the lack of forward and backward linkages. The linkages were often established but in the home location of foreign-owned firms. What made matters worse was the fact that the rent obtained by the country's elites was mostly spent on luxury items or was reinvested in the economies of the core countries. According to the dependency theory, this situation will likely stay unchanged since governments of the mineral-rich states are in an inherently weak bargaining position with large foreign-owned firms who 'have superior knowledge and bargaining power by virtue of their control of savings, investment, technology, and markets' (Gunton 2003: 70). Thus developing countries are locked in the cycle of permanent underdevelopment. This view was reinforced in the 1960s and 1970s by the world-system theory.

The world-system theory, the US version of Latin American dependency theory, provided a highly political approach to the problem of economic development in the third world (Chirot and Hall 1982). Immanuel Wallerstein (1974) argued that uneven development is one of capitalism's characteristics. The core–periphery dichotomy was not unique to the experience between Latin American and the US and could be traced back to the fifteenth-century European feudal economy onwards. From the beginning of the modern age well-developed towns (the core at the time) needed peripheries from which to extract the surplus that fuelled expansion and led to the growth of the capitalist order. Key to Wallenstein's analysis were the historical stages of development which differed from the uniform evolutionary constructs of modernization theory. The world-system theory, as opposed to classical approaches, also took colonialism seriously.

The rise and crisis of the third-world state (1970s–80s)

The first wave of state capitalism

The picture presented by this radical approach was shaken by developments in the Middle East in the 1960s and in Latin America in the 1970s. The formation of OPEC, the nationalization of the oil industry, and the formation of the National Oil Companies (NOCs), by once seemingly dependent states, questioned the claim that peripheries are locked into a never ending cycle of dependency (Marcel 2006; Stevens 2008a; Vivoda 2009). At the heart of these developments was the cooperation between key oil-producing states. Raymond Hinnebusch argues that

> OPEC's ability to engineer massive increase in oil prices, the Arab producers' ability to use oil as a political weapon without suffering military intervention and the accumulation of seemingly enormous 'financial

power' in the hands of the Gulf Cooperation Council (GCC) states seemed to show that the age of imperialism was dead and that dependency had been superseded by interdependence between the core and the oil-producing Middle East.

(2003: 36)

The overwhelming dependency of the US and Europe on Middle Eastern oil until the mid-1970s undeniably put the OPEC countries in a different category to the mineral exporting countries since the global north never relied on minerals from the south to the same extent as Middle Eastern oil. Nevertheless, the nationalization of oil signalled a major shift in the north–south relationships.

Developments in the Latin American minerals sector in the late 1960s further tested some of the assumptions behind dependency theory. Raymond Vernon (1971) crafted the obsolescing bargaining theory (OBT) which critically questioned the idea that governments of the mineral-rich states are in an inherently weak bargaining position vis-à-vis large foreign-owned firms. Vernon argued that governments initially have to strike a relatively unfavourable deal with multinational corporations (MNCs) because of their lack of financial resources, technological capacity and marketing prowess. Thus, at first MNCs are in a much stronger position than the government. However, this situation would not last forever, as dependency theory suggested, particularly in the minerals sector.

Vernon predicted that the conditions underlying the initial host government–MNC bargain would deteriorate – that is, the bargain would 'obsolesce'. As the industry matures and the host government becomes more competent, the relative bargaining power of the companies obsolesces, providing the government with the opportunity to change the terms of the initial agreement. At this juncture, given the large up-front expenditure already committed and the strategic need for ongoing access to resources that generate high rents, investors have very few choices but to work with the host government (Bayulgen 2010: 17–22). The OBT anticipated that multinationals would renegotiate the original contract on less favourable terms or else face expropriation. According to Vernon, the theory would apply increasingly in the future and he confidently predicted that host-country government bargains would continue to deteriorate 'in the future as in the past' (Vernon 1985: 257).

Not all countries in the 1970s limited themselves to merely renegotiating contracts with the MNCs. Some went one step forward and nationalized their extractive industries completely. Governments from various parts of North Africa and Latin America (Algeria, Brazil, Colombia, Ethiopia, Libya, Mexico, Peru and Venezuela) nationalized their oil, steel production and petrochemical industries earning themselves the label of governments pursuing state capitalism (Farsoun 1975; Fernandez and Ocampo 1975; Bamat

1977; Petras 1977). The state's dominant role in the economy of these 'state capitalist' countries also expressed itself through the centralization of finance and banking. Third-world state capitalism was designed, according to its advocates, to aid the development of a strong and viable industrial bourgeoisie, the rise of which had been constrained by the historical effects of colonial and imperial domination.

On this analysis, global capital and the former imperial masters had perpetuated their influence through favouring the growth of a landed and commercial bourgeoisie primarily based on production for export (such as minerals). On a structural and geopolitical level the establishment of an 'alternative' state structure was made possible due to 'the greater intensification of inter-imperialist rivalries and the consequent relative decline of United States hegemony' (Dupuy and Truchil 1979: 2). At the same time, it was claimed that the US, Western Europe and Japan and their dominant classes were not alarmed by the emergence of state capitalist regimes. This was for one fundamental reason. The nationalizations of the 1970s did not lessen these countries' dependence on foreign capital and technology, know-how, machinery and servicing which continued to be required for the basic functioning of those industries. It was concluded that at the end of the day state capitalism 'merely represented a shift in the nature of the relationship between advanced capitalist countries and the Third World' rather than a real structural change (1979: 5).

The rise of the third-world state seemed to rebalance, at least to some extent, the relationship between producing and consuming countries. This rebalancing was partially an outcome a counter-strategy, aimed at underdevelopment, which argued for a greater role of the third-world state in the economic affairs. In order to break the pattern that perpetuated the web of dependencies on a global level advocates of the 'New International Economic Order' (NIEO) and 'Import Substitution Industrialization' (ISI) began regarding the state as the main actor in bringing about economic development, operating both as the formulator of planning objectives and as the executor of industrialization policies. In this equation, mining 'was intended to provide financial resources for development and these were considered to be a "springboard for industrialisation"' (Bastida 2008: 104). Thus, bringing resources under the control of the state was seen as essential for further progress.

The increased involvement of the third-world states in their extractive industries was enabled by greater collaboration between these states. This was particularly true in the Middle East which with the developments in the 1970s became one of the key regions for investigation by political economists. Consequently, between the 1980s and late 1990s there is a clear growth in scholarly studies focused on the oil-rich countries in North Africa and the Middle East as opposed to South America and the mining industry. One of the central features of this new research was a shift in focus from analysis of the global inequalities engendered by dependency to the interaction between oil rents and state structures among the oil-rich states.

In the case of the Middle East, the focus on the state also aimed at redefining our understanding of political processes in the region by showing that 'the impulse towards authoritarianism lies not in something primordial in Arab culture, but instead in more complex dynamics involving economic growth and stagnation, social-structural transformation, state formation and institutions inertia, and ideological transformation' (Crystal 1994: 264). Finally, the focus on the state in the Middle East as well as other non-Western regions of the world was an expression of a larger shift in social science, 'Bringing the state back in' was a rallying cry for those wanting to treat the state as a central explanatory variable and who were highly critical of earlier pluralist approaches in which the state became merely the arena in which social conflicts were waged (Skocpol 1985).

Rentier state theory and resources wars

The oil rent which flowed to the coffers of the oil-producing states from the mid-1970s onwards altered the economic structure of the Middle East. The postcolonial Middle East state, which in terms of economic and political performance did not differ all that much from other postcolonial states in the 1950s and 1960s, began following a very different trajectory in the 1970s and 1980s (Ross 2012: 7). This phenomenon was first analysed within the framework of the Rentier State Theory (RST). The RST attempted to account for the impact that rent, derived especially from the sale of oil and gas on international markets, had on the nature of the political systems of the resource- and energy-rich countries. The proponents of the RST focused on states whose economies were dominated by rents rather than by productive enterprises like agriculture and manufacturing, and where the origins of the income were external. In addition, the rent was generated by a small elite, the majority being only involved in the distribution or utilization of it, and with the state itself being the principal recipient of these rents. The rentier state accordingly played a central role in distributing wealth to the population (Beblawi 1987: 51–3).

One of the key themes in RST was state autonomy. The argument was that major energy exporters were financially autonomous from their citizenry. This was due to the fact that a rentier state did not need to extract surpluses from the local population and the basis for survival was the rent income whose origins were external (Luciani 1987: 69). Although oil rent was not the only income, it certainly predominated in state budgets (Okruhlik 1999: 295). Due to the external nature of the state income, the rentier states were only to a small degree dependent on the production processes of their domestic economies. In effect, the state became financially independent of domestic production groups. The inputs from the local economies, other than raw materials, were insignificant (Mahdavy 1970: 429; Luciani 1987: 69). Lisa Anderson argues that 'virtually no state in the region relies solely on its domestic population for resources, and many governments are often accountable for their spending, when they are accountable at all, to foreign lenders and donors rather than to their own people' (1987: 14). The external nature of rents and the isolation and autonomy that it implies has had far-reaching repercussions, including the lack of a coherent economic policy, the presence of a largely inefficient bureaucracy, the lack of political freedom and the decline of agriculture and industry.

A widely discussed outcome of rentier state's autonomy is the dwindling capacity of its extracting institutions. Taxation, which is the most essential function of the modern state, lost its importance in the eyes of such rulers since they ultimately depend on non-tax revenues (Chaudhry 1989: 103). Thus, the construction of a coherent tax system did not take place or was stopped in its tracks in most rentier states (Tilly 1992; Schwarz 2008). The lack of sufficient extractive institutions results in stripping the state of the mechanisms through which it can gather the necessary information required for the formulation of a coherent economic policy (Ross 1999: 313). In such a situation, political rather than economic considerations become the basis for any sort of decisions (Shambayati 1994: 309).

The lack of sufficient extracting institutions did not mean that bureaucratic institutions in the rentier state countries failed to develop. On the contrary, one of the characteristic features of rentier states was the huge state bureaucracy. Autonomous states that were based on external capital needed extensive bureaucratic apparatuses which would distribute oil revenues in a politically advantageous fashion (Moore 2001: 129). Nazih Ayubi argued that in rentier states 'bureaucracies are expanded in order to provide the ruler with a "stability platform", a control device and a space for extending patronage' (2001: 308). The bureaucracy, whose main role is distribution, became highly undifferentiated and inflexible.

Rentier state theory also argued that since state decision-makers in oil-rich countries were much less constrained by the interests of domestic actors, this resulted in a strong tendency towards authoritarianism (Brynen 1992: 74). Pauline Jones Luong argues that 'natural resources wealth is characteristically found in tandem with nondemocratic political systems' (2000: 28; see also Luciani 1987: 74). According to rentier state theorists, in non-rentier states taxation serves as a lever for society to exercise some political influence over state leaders. Anderson points out that taxation 'binds the populace to the state by creating expectations among taxpayers what they are to receive in return for their contribution to the upkeep of the administration' (1987: 9). In a rentier state, where the state becomes increasingly independent from the society for its economic and political survival, such levers do not exist and petrodollars can simply be used to subsidize consumption, such as for fuel, housing, public services and utilities.

At the same time, spending by the rentier states has produced some notable achievements. These are in increased employment opportunities, generous pension plans and improved public welfare. Terry Karl argued that '[as] a group, the OPEC nations showed a dramatic expansion in infrastructure and allocated a larger share of their national income to education and health than any other group of developing countries' (2000: 36). Yet, other areas of the economy were largely ignored. Most importantly, the Middle East rentier states neglected agriculture and industry. Hootan Shambayati notes that 'the availability of external rents and the fact that the state's income is not determined by domestic production means that the rentier states did not need to increase domestic production' (1994: 309). The rentier states also quickly became victims of 'Dutch Disease', where oil windfalls did not benefit the larger economy but rather ended up hurting the non-oil sector as petrodollars pushed up the real exchange rate of a country's currency and thus rendered most other exports uncompetitive.

The weakness of the rentier states in the Middle East became apparent in the mid-1980s after the collapse of the oil price and the subsequent debt crises (Frieden 2007: 370). Their decline forced rentier states to rethink their strategies in line with the new policies advocated by the international financial institutions and the proponents of structural adjustment programmes (SAPs). The average SAP was premised on 'the privatisation of state companies, the end to state monopolies, the end to subsidies for consumer goods, the devaluation of the local currency and a general commitments to budgetary restraint and macroeconomic stability' (Soares de Oliveira 2007a: 44). In the Middle East, the implementation of some of the SAP principles led to a readjustment of the states' welfare functions and opened the Middle East's oil sector to private players (Schwarz 2008: 602). Within a few years, with the exception of Saudi Arabia, all countries in the region began to allow foreign entry upstream which until then had been off-limits to the International Oil Companies (Stevens 2008b: 35). By the end of the 1990s, rentier states in the Middle East were plagued with 'double-digit inflation, cost overruns in poorly designed projects, and a monumental waste of resources' (Karl 2000: 35).

The crises of the rentier state in the Middle East in this period brought about economic hardship and the need for economic adjustment, but not breakdown of the state structure or a major political upheaval. The situation looked very different in sub-Saharan Africa which towards the end of the Cold War saw a collapse and disintegration of neo-patrimonial regimes (Bratton and van de Walle 1994), a regime type that displayed strong similarities to rentier states (Le Billion 2001, 2004, 2012; Kaldor et al. 2007). Resources, most importantly diamonds, became one of the leading characteristics of those conflicts (Ross 2004, 2006; Reno 2011).

It is today widely acknowledged that the first salvo in the debate linking civil wars to natural resources was 'fired' by development economists Paul

Collier and Anke Hoeffler. Collier and Hoeffler used a comprehensive list of civil wars produced at the University of Michigan and matched this against a mass of socio-economic data, country by country and year by year, with 'the goal of trying to determine the factors that affected the likelihood of civil wars developing in a given country within the next five years' (Collier 2008: 18). Through their large-N study they found that countries which suffer from low income and slow growth, and are dependent on primary commodities, are more prone to civil wars (Collier and Hoeffler 1998, 2002, 2004). Similar conclusions were also reached by other scholars who, each using unique data sets, argued that oil-exporting states are likely to suffer from civil wars (de Soysa 2002; Fearon and David 2003). The United Nations Environment Programme in their 2009 report likewise stated that

> [s]ince 1990 at least eighteen violent conflicts have been fuelled by the exploitation of natural resources. In fact, recent research suggests that over the last sixty years at least forty percent of all intrastate conflicts have a link to natural resources. Civil wars such as those in Liberia, Angola and the Democratic Republic of Congo have centred on 'high-value' resources like timber, diamonds, gold, minerals and oil.

> (UNEP 2009)

According to Collier and Hoeffler, civil wars can be modelled as 'loot-seeking' wars (conflict driven by greed), or as 'justice-seeking' wars (conflicts driven by grievance). The greed-driven 'loot-seeking' wars assume that the principal driver for a (young) person to join a rebel movement is the expected utility of their actions 'which is a function of opportunities foregone by engaging in violence and the availability of lootable income, or the payoff' (de Soysa 2002: 397). Collier later commented that joining the rebellion 'is a bit like joining drug gangs in the United States' (2008: 21). The second model is of 'justice-seeking wars', where groups form with the goal of overthrowing the ruling regime or government and replacing it with a more just and equitable political system. Collier and Hoeffler concluded that greed is a far stronger trigger of conflict than grievance, primarily because the '[c]osts of organizing to fight determine the outbreak of violence and the supply of justice. To put it succinctly, there are fewer martyrs than opportunists!' (de Soysa 2002: 398; for further discussion on the issue, see Ostrowski 2010).

By the mid-1990s the multiple crises of the rentier state, the prevalence of resource-fuelled civil wars and the more general cyclical economic crises of the resource-rich states (Auty 1990) began to be viewed as part of a larger phenomenon which was named the 'resource curse' (Karl 2007). The concept of 'resource curse' served to identify and popularize the idea that there existed a critical economic and political problem which needed addressing in an increasingly globalizing world. Emboldened by the collapse of communist regimes and the rise of the Asian Tigers, global financial institutions such as the IMF and the World Bank saw resource-rich countries as another frontier to be crossed (Rosser 2006: 7). The collapse of commodity prices and the declining power of the third-world states in the 1990s made this look both urgent and attainable.

Globalization and resurrection of the state (1990s–2010s)

The resource curse and its remedies

Proponents of the resource curse thesis argued that, as noted above, historically natural resources have been seen as a blessing for developing countries, as seen in the dominance of the staple and modernization theories of the 1950s. However, they also argued that this conventional understanding of the link between resources and development had not been properly scrutinized. This lack of intellectual interrogation became evident once research began to demonstrate that the abundance of natural resources was actually negatively correlated with economic growth (Sachs and Warner 1995; Auty 2001). At the same time, dependency theory, a theory which could be seen as a precursor questioning the positive impact of 'blessing' of resource wealth was viewed to be too ideological and non-scientific. Drawing largely on the rentier state and Dutch Disease theories, the resource curse school stated that 'natural resources abundance (or at least an abundance of particular types of natural resources) increases the likelihood that countries will experience negative economic, political and social outcomes including poor economic performance, low levels of democracy, and civil war' (Rosser 2006: 7).

The negative impact of resources on a country's development was seen as a global phenomenon. However, the 'ground zero' of the resource curse was sub-Saharan Africa. Central Asia and the Caucasus also became part of the resource curse analysis after the post-Soviet space was cast as another region of the world that could potentially follow the footsteps of sub-Saharan Africa (Ebel 2003). The resource curse proponents adopted an inclusive approach to the issue of the different types of resources. The phenomenon that they talked about had primarily originated in studies that looked at oil but that did not prevent them in extending their analysis to minerals or gas.

The fact that different resources were extracted by very different types of companies or that oil rent differed from the rent generated by the mining sector was rarely seen as a factor worth taking into consideration. Also, the performance of countries that overall benefitted from their resources (Botswana, Indonesia, Chile, Norway, Australia, the US, Canada and Malaysia) was not always properly acknowledged, as it complicated an otherwise fairly clear-cut picture (Walker 2001; Stevens 2003a; Rosser 2006). Furthermore, the issue of ownership structure of the oil and gas sectors also did not feature in the resource curse debate. Ownership was treated as a constant across time and space (Luong and Weinthal 2010).

In general terms, the resource curse thesis was not as significant an intellectual breakthrough as it is sometimes presented to be and its shortcomings are a subject of ongoing debates. However, the advancement of intellectual knowledge was never as important for the resource curse proponents as their policy ambitions to generate policies that would help eradicate the curse once and for all. Hence, it was assumed that we know enough about the negative side of the resources to start putting forward different remedies. It was assumed that in the process the 'curing' of the 'curse' of natural resources would become a source of cooperation rather than conflict.

In the era of globalization, two sets of agents, the World Bank and IMF on the one side, and NGOs on the other, began proposing policies to address the ills associated with the resource curse. The efforts undertaken by global financial institutions produced a voluminous policy literature that proposed a whole host of actions that should be taken by the resource-rich states and the resource-extracting companies (Humphreys et al. 2007) as well as concrete projects on the ground such as the Chad-Cameroon pipeline project (Soares de Oliveira 2007a). The NGO-driven initiatives took a slightly different approach to the one favoured by the global institutions and firmly fixed their eye on the problem of corruption. Their key area of concern lay with revenues and revenue losses resulting from opaque practices.

Instrumental in the rise of the corruption agenda were reports published towards the end of the 1990s by the London-based NGO Global Witness that dealt with Angola's 'missing billions' as well as the links between conflict and the exploitation of natural resources. Another important building block was the 'Elf Affair', in the beginning of 2000s, which exposed 'a tangled web of political ambitions, influence peddling, oil, and corruption' at the heart of the French oil industry (McPherson and MacSearraigh 2007: 200; see also Shaxson 2007; Silverstein 2009). The outcome of those growing concerns was a formation of NGO alliances and the emergence of Publish What you Pay (PWYP) campaign which sought to have companies publish their payments in order to increase transparency in resource-rich countries. To achieve its goal, PWYP pushed for mandatory disclosure of revenue payments by companies to the governments of oil or mineral-rich states. PWYP left a considerable imprint. It main success came when its ideas were picked up by major financial institutions and key Western politicians of the time.

In 2002, UK Prime Minister Tony Blair launched the Extractive Industries Transparency Initiative (EITI), a multidonor trust fund run by the World Bank. The EITI is a coalition of governments, companies, civil society groups, investors and international organizations. Its main objective is to strengthen governance by improving transparency and accountability in the extractives sector. The EITI supports improved governance in resource-rich countries through the verification and full publication of company payments and government revenues from oil, gas and mining.[2] The UN General Assembly unanimously adopted a resolution in September 2008 backing the EITI, and numerous other strongly supportive statements have emerged from the G8 and several other bodies. The EITI is a voluntary initiative that countries sign up to; companies do not sign up but they can be official supporters.

The success of the EITI has been mixed as the voluntary initiative was effectively diluted by companies and resource-rich states (Shaxson 2009). The long-term durability of EITI as well as other policies put forward by resource curse proponents were challenged by a number of developments in the twenty-first century. The rise of oil and other commodity prices, the financial and economic crisis, the rise of new powers and the subsequent declining power of the West have had significant repercussions on the prospects for reconciling resource extraction with development and economic growth. As the authors of one study remark,

> it is important to appreciate the conditions that enabled the rise of the reform agenda: a unique window of opportunity of unquestioned Western dominance in world affairs and energy markets, lasting from 1990 until 2005, set the normative tone (if not the substance) of the post-cold war international system. The rise of China and India, the arguably diminished status of the West, and the rise of a plurality of global power centres means that this window for normative tone setting may have now closed.
>
> (Benner; Soares de Oliveira and Kalinke 2010: 294)

This rise of the new non-Western players led to a shift in the field of political economy. Since the mid-2000s the debate had been increasingly dominated by the looming spectre of the state and its return. The BRIC countries led the pack and the concept of state capitalism began working its way back into the analysis. Resources once more came to be seen as a source of possible conflict and tension on the global level, whereas voices critical of resource-based development have been gradually suppressed.

The second wave of state capitalism

Ian Bremmer, who today is the most vocal and well-known proponent of the revived state capitalism concept, argues that state capitalism is a system where the state is the principal actor in the economy. Politicians use markets for ultimately political gains and there is no rule of law. State capitalism is portrayed as a system which presents a real alternative, or more accurately, a challenge to the global free-market system. The current version of state capitalism first took root among authoritarian and non-Western regimes, the key states being China, Russia and Saudi Arabia, which are all important players in international politics and the global economy. In all three states the ultimate motive of the ruling elites, according to Bremmer, 'is not economic (maximizing growth) but *political* (maximizing the state's power and the leadership's chances of survival)' (2010: 5). Another author argues that state capitalism 'is ultimately a story about economic nationalism and global power politics, especially when it comes to energy, food and logistics' (King 2010: 150). It is important to note that the notion of state capitalism does not have the ambition to be a catch-all concept that can be applied to all parts of the world to the same degree. It is also not a coherent political ideology. Rather, it is a set of principles that governments can and increasingly are adapting to meet their particular needs.

The embodiment of state capitalism on the international stage is a new class of companies that have pushed their way onto the international arena in recent years. The key feature of those enterprises is the fact that they are either owned or closely aligned with their government. Thus they are not multinational corporations in the traditional sense of the term, since they answer first and foremost to politicians rather than to shareholders. On the domestic front regimes that champion state capitalism have made sure that strategic national assets 'remain in state hands and that governments maintain enough leverage within their domestic economies to safeguard their survival' (Bremmer 2010: 21). In short, such control of key industries is supposed to both guarantee political stability and deliver controlled and sustainable growth.

Whereas the term 'state capitalism' is not new, its current incarnation is distinctive. In a well-known story – articulated in both Francis Fukuyama's *End of History* and Samuel Huntington's *Third Wave* – the collapse of the Soviet Union and the end of the Cold War marked a victory for the free-market economy over socialism and democracy over authoritarianism. It was not assumed that the way forward would be paved with roses but that nevertheless the world was moving together towards a new democratic order. However, this did not happen. Rather, authoritarian regimes remained authoritarian with the notable exception of South America and Central and Eastern Europe and, in the majority of cases, learnt to compete internationally by embracing market-driven capitalism. Thus, what emerged from the ashes of the 1990s has been a hybrid type of authoritarian regime: liberal in an economic but not political sense. Eventually, the conviction in the inevitability of democracy and the success of the 'third wave' of democratization began to fade.

Initially, the embrace of market-driven capitalism was primarily due to a widespread belief that command economies were doomed to fail. At the same time, it was clear that market forces left entirely to themselves present real political problems. The free market could spin out of control, enabling financially powerful political players to challenge the status quo. The turning point in the rise of state capitalism beyond its 'founding members' was the financial crises of 2008. The market meltdown reversed a move towards less government intervention in the West and 'discredited free-market capitalism for many in the developing world' (Bremmer 2010: 46). Another important factor in the growth of state capitalism was the economic success of its key promoters, which has attracted imitators throughout much of the developing world. This led to a situation in which we cannot talk any longer about one type of state capitalism but rather of many variants of it. Today, state capitalism has many shades of grey and is not exclusively associated with authoritarian regimes any longer. In order to manage state capitalism, politicians use a whole host of intermediary institutions, 'tools' over which a state can, but does not have, to exert day-to-day control. The most important are National Oil – and Gas – Companies, other State-Owned Enterprises (SOEs), privately owned national champions, and Sovereign Wealth Funds (SWFs).

The rise of state capitalist states in the 2000s as in the 1970s is largely but not exclusively associated with the natural resources. Yet, there are at least two important differences between today's version and the earlier one. First, countries that engage in state capitalism are not peripheral to the global capitalist economy but are increasingly vital to its very functioning (Hart and Jones 2010/2011). Second, thanks to SWFs the states in question and their indigenous bourgeoisie are no longer 'dependent' on foreign capital to the same extent as they once were so as to develop and finance their extractive industries. What is more, the technology once available exclusively to the companies from the global north can be now purchased on the international markets. Much of this technology resides with engineering contractors and service companies which, in partnership with NOCs, are just as well placed as IOCs to invest in R&D to develop new technologies.

The next steps for research

The rise of state capitalism presents a new chapter in the story of the political economy of fossil fuels and minerals. Yet, this does not mean that the problems associated with the resource-based development are likely to disappear or that the tools used to analyse resource-rich states will stop being useful. The existing record testifies to the contrary. In recent years the rentier state theory, besides the Middle East region, has been applied to other areas of the world, including sub-Saharan Africa (Yates 1996; Clark 1997; Frynas 2004; Soares de Oliveira 2007a), South America (Karl 1997) and more recently to post-Soviet Central Asia (Ishiyama 2002; Kuru 2002; Franke *et al.* 2009; Ostrowski 2011a) and Russia (Luong 2000; Kim 2003; Rudiger 2005; Goorha 2006; Tompson 2006; Wood 2007). At the same time, existing criticism of resource curse studies and the emergence of state capitalism make clear the need for a new research project that looks much more closely at questions of development in connection to companies that are vital to state capitalism, most importantly, the National Oil Companies and State-Owned Enterprises. In other words, the extracting industry, which was missing from the political economy analysis since the 1970s, should be brought to the forefront of analysis (Soares de Oliveira 2007b; Bridge 2008a; Downs 2008; Stevens 2008d; Losman 2010; Obi 2010; Mitchell 2011).

We still know remarkably little about the NOCs and other national champions – which are by far the most powerful companies in the oil-rich countries – and the political struggles that they generate. Students of rentierism often assume that the president or ruler and his family control those companies with great ease, completely ignoring the political battles between elites in the NOCs and the ruling regime. Yet, if we were to scrutinize – over a long period of time – the political struggle for the control of those companies and the larger oil or gas complex in the resources-rich state, we would arguably begin thinking very differently about the forces and actions that shape the political reality, the state structures and economic development in many of those countries. It is important to keep in mind that the type of political struggles and their logic have their origins in the very nature of the oil and gas industry (Ostrowski 2011b).

Conclusion

The political economy of the resources has undergone a remarkable development since its inception in the beginning of the twentieth century. It has generated substantive and extensive academic debates that cyclically has managed to spill over and inform wider political debates. Indeed, how we imagine and think about the history of Latin America in the 1950s and 1960s, the Middle East in the 1970s and 1980s and sub-Saharan Africa in the 1990s and 2000s is very strongly linked to debates that dealt with the issues of resources and development. Yet, those debates did not unfold in a vacuum. The events and discourses that took shape in a particular period or regime also influenced ways of thinking about the political economy of resources. This chapter demonstrated that all those multiple factors have to be taken into consideration in order to understand why today, as in the past, resources can bring about conflict and/or cooperation.

Notes

1. 'World War I', *U.S. Geological Survey 1879–1989, U.S. Department of the Interior*, http://pubs.usgs.gov/circular/c1050/ww1.htm [Accessed 5 January 2012].
2. http://eiti.org/ [Accessed 15 March 2012].

Part III

Companies, Contracts and Communities

6
Corporations vs. States in the Shaping of Global Oil Regimes

Giacomo Luciani

The tension between the sovereign state and the multinational corporation has long been in focus. Globalization – the progressive opening of markets to the free circulation of goods, services and capitals (not necessarily individuals) – has been perceived as limiting the power and sovereignty of states. In a globalized world, governments are increasingly obliged to compete with each other to attract investment and foster the competitiveness of their industry through progressive deregulation, that is, simplification or outright abolition of all sorts of rules that states may have imposed over time to pursue various objectives deemed in the collective interest. Through competition, states may be expected to progressively self-limit their intervention tools and their penetration of economy and society, leaving a free hand to market forces and corporate actors.

Attempts at 'governing' globalization – that is, agreeing on a set of rules to impose on all states and corporations – rarely yielded concrete results, except in the direction of dismantling existing limitations on trade and investment. In many instances, the only acceptable outcome has been industry 'self-regulation', for fear than anything imposed from the top by even a powerful set of governments may end up damaging their competitiveness vis-à-vis outside 'free riders'. States are expected to guarantee law and order and contribute to enforcing contracts, not much more. States are certainly not expected to directly engage in any economic activity, and privatization of any state-owned enterprise is considered a must and beyond doubt.

With respect to the oil and gas industry, globalization led to an expectation that 'resource nationalism' would inevitably fade away, and companies would gain untrammelled access to resources everywhere – reversing the tendency to state control of natural resources that progressively gained force in the third quarter of the twentieth century (1950–75). However, the globalization narrative never provided an adequate fit to the reality of the oil and gas industry.

This chapter focuses on the dialectic between corporations and states in the definition of – or absence of – a regime for the global oil and gas industry. The existence of a regime – a set of written or unwritten norms that most relevant actors abide by, and those that do not face some kind of additional cost – is the key to the predominance of cooperation. In the absence of a regime, conflict prevails. The chapter argues that in the oil and gas industry states have always been directly involved with corporations – their creation, ownership structure and investment decisions. Yet the most successful regime in history was established by companies and at least officially hidden to the eyes of the states. The latter saw the outcome and they liked it, so acquiesced. That regime was undone by the decolonization process and the new assertiveness of the producing countries' states, which in turn established their companies. The producing countries established an OPEC regime – state controlled and multilateral – which however was never universally accepted and soon collapsed. The subsequent regime, based on self-regulated markets and progressive financialization, has not brought about the expected retreat of states. In fact, it has generated growing price volatility, and is leading to a comeback of the states in the context of an emerging state capitalist regime – that is, close association between state and companies and bilateralization of relations – in an attempt to reduce uncertainty and recurrent shocks.

Birth of companies: State-supported or independent?

Oil companies have in many cases been supported by their respective national governments right from their establishment or very soon thereafter. This has been specifically the case with respect to the discovery and exploitation of oil in the Middle East. In contrast, the US government was little involved in the birth of domestic oil companies. Elsewhere in the early oil-producing countries – in the western hemisphere (Mexico, Venezuela) or indeed in the Russian empire (Baku) before the Bolshevik revolution (Stevens, Chapter 1, this volume) – active involvement from the government of the companies' country of origin was less evident.

It is an interesting historical question exactly why the governments of the major industrial powers of the time took such acute and direct interest in Middle East oil. Where the resource was available within the boundaries of the interested state – such as in the US – government's emphasis was rather laid on maintaining a competitive environment among a plurality of independent private players. This is demonstrated by the US attitude towards Standard Oil, culminating in the break-up of the company in 1911 (see Chapter 1). Yet, in that occasion all international participations, as well as the right to the brand name outside of the US, were attributed to just one of the spin-off companies – the Standard Oil Company of New Jersey – pointing to the fact that different logics were at play domestically and internationally.

In Venezuela and Mexico local dictators saw no problem with the domination of foreign companies – which in any case was not monopolistic, in the sense that control of resources was distributed among several independent companies. In Russia, national companies developed alongside foreign interests. The Middle East was different. Governments became deeply and directly involved in the activities of the oil companies, and a political logic was on display from the beginning.

In his capacity as First Lord of the Admiralty, Winston Churchill succeeded in 1914 in getting the British Parliament to approve the acquisition of a majority interest in the Anglo-Persian Oil Company, which had discovered oil a few years earlier (Yergin 1991: 156–64). This development inaugurated an era in which governments took keen and close interest in the fortunes of their respective oil companies. Churchill's speech in parliament (Rhodes 1974: 2309) attacked the combined monopoly of Standard Oil in the US and Royal-Dutch/Shell in the 'Old World', and asserted that Anglo-Persian was one of the very few independent companies which may allow a major customer, such as the Admiralty was becoming, to escape inflated prices. He also argued that, considering that the Admiralty contract would greatly increase the value of the company receiving it, it would be logical for the government to acquire majority participation, so it would also reap the benefits. Such arguments became very common in the following half-century, and retained traction until the late 1960s, becoming the standard justification for almost all European governments to establish national oil companies.

The peculiarity of the Anglo-Persian concession was of course that it covered the entire territory of Persia, depriving the government of any alternative interlocutor. In the eyes of the Government of Persia, the fact that the company it had to deal with was majority-owned by the British government had clear implications.

Next came Iraq. There, political interests and preoccupations figured prominently in the mind of the 'architect' of the Iraq Petroleum Company, Calouste Gulbenkian. Even before the First World War and the demise of the Ottoman Empire, he had been busy establishing the Turkish Oil Company to exploit oil deposits in the Mosul vilayet, and called on potential investors with an eye to maintaining political equilibria. It would be an exaggeration to say that the Allied thrust against the Ottoman Empire was primarily motivated by oil, but the importance of the latter became evident in the transition from the original Sykes-Picot agreement, which assigned Mosul to France, to the final Sanremo agreement, when Mosul was added to the vilayets of Baghdad and Basrah to form Iraq, under British protectorate (Shwardan 1985).

Thereafter, Gulbenkian moved to incorporate all dominant Allied interests in the new formation of the Iraq Petroleum Company, whose equity was split in five parts: four equal parts of 23.75 per cent each to Anglo-Persian, Royal Dutch-Shell, Compagnie Française des Pétroles and two Standard off-shoots: Standard Oil of New Jersey and Standard Oil of New York. The remaining 5 per cent remained in the hands of Gulbenkian's company, Partex (Yergin 1991: 202–3). It should be noted that the Compagnie Française des Pétroles was created by col. Ernest Mercier, following the suggestion of the French Prime Minister Raymond Poincaré, essentially in order to take up the French share in the Iraq Petroleum Company – the same share which in the original Turkish Petroleum Company set-up had been assigned to Deutsche Bank. IPC was granted a concession, after protracted negotiations with the Iraqi side (Penrose and Penrose 1978: 56–74). Iraq was granted full independence only after the concession was signed. Throughout the process, the direct involvement of governments was clearly visible.

Having formed IPC, Gulbenkian insisted that participants in the consortium sign what became known as the Red Line Agreement (Yergin 1991: 203–6). The red line was drawn on a map to define the territories formerly under the sovereignty of the Ottoman Empire, and the agreement stated that participants in the IPC consortium pledged to be involved in the exploitation of any oil to be discovered within the red line exclusively through consortia with the same composition as the IPC. Hence, if one of the IPC consortium members were to discover any oil or obtain a concession elsewhere within the red line, it would have to offer this asset to the remaining members in the same 'geometry' as in the IPC.

This supremely important agreement had far-reaching consequences and was at the heart of the 'companies' regime', as will be discussed in the next paragraph. It certainly envisaged a structure that would force all governments in the former Ottoman Empire to deal exclusively with just one interlocutor, a consortium shaped by politics and very clearly supported by three dominant countries: Britain, France and the US.

As Kuwait was not included in the red line, the granting of a concession there was hotly contested between the Anglo-Persian Oil Company and Gulf Oil of the US. Although Kuwait was not fully independent, the Amir successfully pitted one company against the other, and finally granted the concession to a 50/50 joint venture of the two – thus narrowly escaping total British domination, but still having to accept a single interlocutor. When oil concessions were under discussion in Bahrain and Saudi Arabia (Yergin 1991: 280–92), IPC was offered to acquire them, but it either was not interested (in Bahrain) or bid lower than the competition and lost (in Saudi Arabia – to Standard Oil of California, that was not in IPC). Elsewhere in the Ottoman Empire, IPC lookalikes ruled.

The granting of a concession in Saudi Arabia did not lead to an immediate discovery. The concession was given to Standard California in 1933, but oil was discovered only in 1938. Standard California was joined by Texaco in a 50/50 joint venture (Arabian American Oil Company, ARAMCO), but the Second World War prevented the speedy development of the discovery. After the war, the two companies, taken aback by the huge dimension of the project, looked for additional partners, and closely involved the US government in the search (Anderson 1981: 56–67). Indeed, at some point the Truman administration even considered taking over ARAMCO tout court, following in Churchill's steps and even surpassing him. Eventually this was deemed politically unacceptable, and the final shape of ARAMCO was defined in 1948, when Jersey Standard and New York Standard stepped in, with a participation of 30 and 10 per cent respectively (leaving California and Texaco with 30 per cent each). Direct government involvement was thus very evident in the birth of ARAMCO as well as of the other main companies in the region.

The intense government interference in the organization of the industry in the Middle East is probably a reflection of shifting equilibria that created an unstable political environment. The Ottoman Empire had to be partitioned, the division between France and the UK was not well defined until Sanremo, and was not destined to last long; the US was the ascending power and needed to be both contained and accommodated. No such issues existed in Venezuela or Mexico. The uncertainties of the political framework within which the industry was to develop, together with the perceived 'strategic' nature of the commodity, may be said to have determined the politization of the industry itself.

The companies' regime and state support

There was, however, no grand political agreement to create a public international governance of the oil industry. The definition of a 'regime' was left to the companies, and was based on a set of interlocking private contracts and agreements. There was, first of all, the composition of the IPC, in which five out of the eight dominant oil companies (which were later dubbed the seven or eight sisters,[1] depending on whether the smallest of them, CFP, was included or not) were present. The IPC was not a joint venture – producing and marketing oil on its own account – but a service structure, producing oil on the account of its members and making it available to them in proportion to their equity participation. The decision concerning how much oil would be produced each year was taken by a qualified majority according to a rule that allowed companies with abundant supplies to limit production increases (Penrose 1968: 154–71). 'Long' companies – notably Anglo-Persian, which controlled also the whole of Iran's output and 50 per cent of Kuwaiti production – agreed to sell some of their oil to 'short' companies (such as notably Royal Dutch Shell) at preferential prices.

The three 'sisters' who were not members of IPC (Gulf, Standard California and Texaco) were connected to the core of five IPC members through their common shareholding in, respectively, the Kuwait Oil Corporation and ARAMCO. Hence, clear mechanisms existed among these eight companies to coordinate production decisions, manage exchanges within the group and stunt competition among themselves as well as from other potential entrants. The sisters controlled some 75 per cent of global oil production and maintained prices stable and competitive, allowing the rapid penetration of oil and displacement of coal, which was still the dominant primary source of energy in the mid-twentieth century. This regime created conditions quite conducive to the success of post-war reconstruction and prosperity in the West.

Governments were not directly involved. In fact, in 1952 the US Senate conducted an inquiry and produced a Staff Report to the Federal Trade Commission entitled 'The International Petroleum Cartel' (United States Federal Trade Commission 1952), which analysed and laid bare the mechanisms through which the sisters jointly governed the industry. However, no action was taken to dismantle the cartel, which we may interpret as a sign of at least tacit connivance. The justification for this attitude is not difficult to find: the sisters' regime prevented a collapse of oil prices, whereby Middle East production would have displaced almost all production from other regions in the world – an outcome that no one in the West really would have welcomed.[2] Prices were kept sufficiently high, allowing for extraordinary profits, which in turn allowed companies to engage in huge investment funded out of company cash flow. Oil production, refining and distribution grew rapidly, because at the same time oil was cheap enough to displace coal and fuel, the West's economic miracle in the 1950s and 1960s. With hindsight, living, as we are, in a period in which price volatility is a serious obstacle to the implementation of required investment projects, and the whole system has become dangerously fragile, it is necessary to admit to the advantages of the sisters' regime: it was indeed the golden age of oil – except that benefits were not 'equitably' distributed.

Conflict with host states and parent state support

The direct involvement of the parent governments was visible in support given to the companies when they either clashed with host government or disagreed with them on key matters. There were innumerable episodes substantiating this claim, but two are especially telling: ARAMCO's decision to accept an increase in income taxes up to 50 per cent and resistance to the Iranian government's nationalization of Anglo-Persian, culminating in the creation of the Iranian Consortium.

The first episode was relatively low-key and has attracted less attention – in essence because it did not lead to open conflict. As oil production expanded, and the enormity of the oilfields discovered by ARAMCO became evident, the Saudi government claimed for an increased share of the benefits, pointing to the precedent of Venezuela. ARAMCO initially resisted, but then approached the US government to seek a way out. The US government accepted that income taxes paid to the Saudi government would be deducted from ARAMCO's tax liability in the US, which was of the order of 50 per cent of profits.[3] This allowed ARAMCO to accept an increase in their Saudi income tax rate up to 50 per cent, leading to the so called 50/50 split of benefits. The agreement amounted, in essence, to a transfer of tax income from the US to the Saudi government, with little impact on the company. Other producing countries immediately claimed the same, but other parent-company governments, notably the UK's, were not as happy to facilitate agreement as the US's. Therefore, the request for a generalization of the 50/50 split was quite contentious and led to significant tension.

The Iranian nationalization, decreed by the government of Mohammad Mossadegh, was in comparison an epic affair, which sealed the power of the sisters for the following 20 years (Yergin 1991: 450–78). Governments were involved at all stages, from fruitless attempts to avoid the crisis, to attempts to mediate during the crisis, to finally organizing the coup that led to the demise of Mossadegh and setting up the Iranian Consortium that took the place of Anglo-Iranian. Key to the conflict was Anglo-Iranian's call for a boycott of Iranian oil, which the other companies in the cartel supported by allowing for an increase in Iraqi, Kuwaiti and Saudi production. Iranian oil exports collapsed to nil, and attempts to buy Iranian oil on the part of independent refiners were thwarted when their cargoes were confiscated when calling at ports under British jurisdiction (such as Aden). The boycott precipitated an economic crisis in Iran, which undermined Mossadegh and demonstrated that the cartel had the tools to enforce compliance with the regime.

The cartel was strengthened by the creation of the Iranian Consortium: after the fall of Mossadegh, the Shah refused to simply rescind the nationalization, and reserves remained the property of the National Iranian Oil Company, but the management of the fields and marketing of the oil was entrusted to a newly created consortium. Anglo-Iranian – rechristened as British Petroleum – retained only a 40 per cent interest in the consortium, and was forced to make room for the remaining sisters in compensation for the support they had extended during the crisis. Thus the consortium included Royal Dutch/Shell with a 14 per cent interest; the five American companies (Esso, previously Jersey Standard, later Exxon; Mobil, previously New York Standard; Gulf; Texaco; and California Standard, later Chevron) with 7 per cent each; Iricon, a consortium of smaller US companies, with 5 per cent; and Compagnie Française des Pétroles (later Total) with 6 per cent.

The episode demonstrated that the cartel was solid and the sisters' regime could not be challenged by an individual producing state, albeit one as important as Iran was. In fact, the cartel was further strengthened and became more visible, in the composition of the consortium.[4]

Competition and the weakening of the companies' regime

In 1960 a group of producing countries established OPEC (Seymour 1980: Chapter 2), but it was not this organization that led to the demise of the sisters' regime. It was rather a combination of market forces and evolving global economic equilibria. Market forces played a role because high profits attracted new entrants. The sisters maintained a tight grip on the resources of all 'old' producing countries, but oil and gas were discovered in new countries one after the other, and new entrants had opportunities to get their 'place in the sun'. New entrants included so-called independents, that is, private oil companies that were not tied to the sisters, and the national oil companies of the importing countries. Most OECD countries found it appropriate to create or support national oil companies to protect the national interest and avoid excessive dependence on the cartel. Even France, although CFP was a French-owned company, decided to establish a state-owned 'régie' alongside – ELF. Besides the US, UK and the Netherlands (home countries of the seven sisters) only the defeated countries of the Second World War – Germany and Japan – never truly established a national oil company. In short, competition for access to reserves significantly intensified.

Governments of the new producing countries saw their bargaining position improve steadily. They were, from the 1950s onwards, more effectively independent than before[5] and had access to better expertise and advice. Also, the potential value of oil was quite evident, and the mistakes of the past were not repeated. Libya was an especially important case because several new companies obtained concessions, and discoveries were important. For several of these companies – notably Occidental and Hunt – their Libyan oil was by far their most important asset. This meant that the government was in a position to play one company against the other and obtain better terms. For as long as the monarchy was in power, the government's negotiating strength was leveraged to push for increased production, which in fact expanded rapidly. However, things changed when Moammar Gaddafi came to power in 1969. The new government decided that it was not interested in expanding production, and in fact moved to unilaterally impose a reduction across the board (Parra 2004: 122–5). This was partly the outcome of debates within OPEC, which had been considering the need to slow down exploitation of resources, in order to extend their lifetime horizon.

It was, quite likely, also connected to the US' fiscal and balance-of-payments crisis, which undermined confidence in the dollar and appetite to hold dollar assets – in exchange for the real asset that was oil in the ground. The US under Nixon was embroiled in the Vietnam War and its global leadership was seriously under stress. In 1950, the British Empire was already in decline, but US economic and political hegemony was unchallenged outside of the Soviet sphere. Things looked quite different 20 years later – especially in the Arab world, where the increasingly pro-Israeli stance of the US and the humiliation of 1967 had taken a heavy toll.

The first company to be requested to cut back production, Occidental, initially attempted to resist Gaddafi's requests and asked Exxon to compensate the loss at cost. But Exxon declined to offer support; thus, Occidental accepted the government's request of increased prices. The Libyan example was closely followed by the Shah in Iran – demonstrating that ideology was not the key motivation. Eventually the action of these two countries led to the signature of the so-called Tehran-Tripoli agreements, which represented a capitulation on the part of the oil companies and signalled the end of the regime. Then a wave of nationalizations took place: Algeria nationalized (51%) the French companies in 1971, Iraq nationalized (100%) IPC in 1972, Libya nationalized (mostly 51%) in 1973 (Seymour 1980: 218–30; Yergin 1991: 583–5, 646–52; Aissaoui 2001). In no case companies were able to stage a boycott like the one that had brought Mossadegh to his knees.

Prices did not reflect the new reality until 1973 and the Yom Kippur War, but it is not the latter that brought the sisters' regime to an end. It had lost effectiveness already three years earlier, and the Yom Kippur War was just the trigger to precipitate higher prices.

The sisters' regime was replaced by a rather short-lived OPEC regime. It is in fact not clear whether an OPEC regime ever truly existed, because OPEC never had the power to influence decisions and investment outside its own borders, in the same way as the sisters had enjoyed. A majority of OPEC member countries asserted their control over national resources by either nationalizing foreign companies or acquiring 50 per cent or more equity in producing joint ventures. Furthermore, OPEC relied on prices that were decided collectively and imposed from above. However in fact prices increased in waves in response to political panics (the Yom Kippur War, the Iranian revolution, the Iraq-Iran War) and OPEC simply attempted to defend and consolidate whatever higher prices had been reached on 'the market' – in essence a ratchet effect. When prices exceeded fundamentals and competition from the rest of the world tended to erode them, OPEC responded with quotas and further reduction in production. But eventually it had to cave in in 1985, signalling the end of the OPEC regime.

Roots of the turnaround: Weakening of states or weakening of companies?

Why is it that the sisters' regime collapsed and could not be substituted for by a stable alternative? The causes certainly are complex and may be described as a combination of the weakening of states and of companies simultaneously. The sisters were very much the product of post-First World War global political equilibria. The composition of the cartel was revised after the Second World War, and American companies became more important than the British, but the decline of Britain as a colonial power greatly reduced the political support available, in particular to British Petroleum. On the US side, 'Big Oil' always was very influential but never quite managed to control the political arena and be identified as an essential component of national interest.

It may be proposed – and has indeed been argued several times – that the major international oil companies had interests not too distant from those of OPEC: after all they too benefitted from higher oil prices – and indeed continue to do so. Thus, some kind of informal condominium might have emerged between OPEC and the international oil companies, eventually establishing a new international petroleum regime. However, the US government chose a very hostile attitude towards OPEC and sponsored the creation of the International Energy Agency as a counter to the producers' organization. Talks about the potential for cooperation and the need to establish a new economic order – as favoured primarily by France – were always dismissed in Washington (Fattouh and van den Linde 2011).

Companies made important discoveries outside of OPEC (van den Linde 1991: 179–93). The 1970s was the time when North Sea and Alaska production was rapidly increasing. In both cases, BP was prominently involved, and might not have survived otherwise. Of the other former sisters, Gulf Oil disappeared almost immediately; Texaco was eventually taken over by Chevron, and Mobil by Exxon. All of them reacted to the loss of upstream assets by trimming down their refining and distribution assets. Exxon and Mobil briefly made attempts at diversification into altogether different industries, being convinced that there might be no future in oil. In short, companies were also weakened.

Conditions were created for the progressive development of markets where previously they had not existed. A good number of companies were involved in the North Sea and no one dominated the scene. This facilitated the establishment of the Brent market (Horsnell and Mabro 1993: 21–8) as a self-regulated spontaneous-birth creature, which became progressively more complex, although it never was truly transparent. In the late 1980s and after several unsuccessful attempts, a futures contract was launched and the reference pricing system came to be adopted by several exporting countries, Brent becoming one of the two key benchmarks (Mabro 2005; Fattouh 2011). In other words, more and more people became convinced that the 'free market' was the solution, and no other regime was needed.

Retreating states, deregulation and the promise of the market – the limits to privatization

In this context, states began retreating from the industry – in parallel with similar developments in other formerly highly regulated industries like telecommunications, transportation or utilities. State-owned companies were progressively privatized in almost all oil-importing countries – although in some (e.g. Austria and Italy) the state retains an important participation in the equity of the former national oil company, and elsewhere the state enjoys a 'golden share' that has the purpose of preventing a possible hostile takeover. Where such precautions have not been adopted, former national oil companies have disappeared, for example, Belgium's Fina and France's Elf have both been taken over by France's Total (also a French company: the French government did move to prevent a possible takeover of Elf by a foreign company, and did the same later for Suez, that was then taken over by Gaz de France). In short, states retreated but not quite 100 per cent.

Even in Thatcher's UK the retreat was not total. The remaining participation that the British government held in the capital of British Petroleum was floated on the market in an IPO in October 1987. Market conditions were not good at the time, but the Iron Lady refused to call the IPO off. This created conditions whereby the Kuwait Investment Office (KIO) was able to pick up 21.6 per cent of BP's equity from the secondary market, thus becoming by far the most important investor. In this case, the market did not quite work in a way that Thatcher found acceptable. So, the case was referred to the Monopolies and Mergers Commission, which investigated it, found that KIO's holding could operate against the public interest and forced KIO to sell its shares in excess of 9.9 per cent of the equity of BP (Monopolies and Mergers Commission 1988). To prevent a collapse in the value of the shares, BP was forced to buy back the excess shares from the KIO at market-related prices. KIO made a good profit, but it was demonstrated that the retreat of the state and reliance on the market had limits: there still are red lines, albeit different from the one that Gulbenkian drew around the borders of the former Ottoman Empire. In the US, similar red lines became apparent in 2005, when the Chinese National Offshore Oil Corporation (CNOOC) launched a bid to acquire UNOCAL: Congress intervened and blocked the deal (Nanto et al. 2005); UNOCAL was taken over by Chevron.

Nevertheless, for the best part of the 1990s and early 2000s it was commonly maintained that 'resource nationalism' would inevitably come to an end and oil-producing countries would be forced to again open up their doors to the international oil companies. The collapse of the Soviet Union eliminated the separation of the Soviet oil industry from the rest of the world, alimenting the conviction that reserves would be made accessible to the international oil companies, and OPEC countries would in turn be forced to capitulate.

Limited openings did in fact take place in many countries, but emphasis should be put on limited. Kazakhstan and Azerbaijan are important examples of plays that were previously closed and opened up. But elsewhere, from Russia to Brazil, international oil companies have been allowed to gain only very limited access to natural resources, and national companies remain dominant. Even in Iraq, notwithstanding the massive US military presence and all the talk about privatizing oil assets and having Iraq quit OPEC, control has remained very much in the hands of the state.

The weakness of companies and the rule of financial intermediaries: The loss of the strategic perspective

International oil companies have been weakened by the competitive environment and increasing influence of financial analysts and intermediaries. For long periods of time it has been cheaper to acquire reserves on the stock market rather than through genuine exploration, discovery and development of new fields. All companies perceive themselves as potential takeover targets and must worry about their price to earnings (P/E) ratio, which has been and continues to be very low. The market values the stock of technology companies even if they turn in losses year after year, and snubs highly profitable oil companies: simply put, they are viewed as not having much of a future.

To survive in this environment, companies must pay attention to the ratings and recommendations of financial intermediaries, and this has led to overemphasis on reducing costs, short-term profitability and returning value to the shareholders. Cutting costs has been achieved through outsourcing of even some highly sophisticated operations and reduction of manpower. Mergers and acquisitions have been a tool to increase the reserve base of the acquiring company and further reduce costs, as normally larger assets can be managed without significantly increasing employment. The industry has destroyed jobs in large waves especially in the 1990s, liberating a vast pool of technical expertise, which previously had not been available outside of the major companies. This in turn facilitated the growth and acquisition of competencies on the part of service companies and national oil companies. Collectively, the international oil companies have engaged in intensive cannibalism. Today's four 'supermajors' are all the result of conquest: BP successively absorbed Sohio, Amoco and Arco, but has not, as intended, been able to absorb TNK-BP; Chevron absorbed Gulf, Texaco and Unocal; Exxon absorbed Mobil. Only Shell made just relatively minor acquisitions.

The value of vertical integration has been questioned, as companies are encouraged to concentrate on the line of business in which they can achieve best results, and divest all unnecessary complements. Smaller companies are frequently pure oil producers, and rely on the market to sell their oil. The role of physical trading companies has been expanding, and so have independent refiners.

Companies have lost much of their strategic perspective and are reluctant to engage in investment whose gestation requires years. There are exceptions to this generalization, but they are not many. The majors find it more and more difficult to add to their reserve base through new discoveries, replacing the oil that they extract year after year. Reserve additions are more frequently the outcome of re-evaluation of old fields rather than genuine new discoveries. Otherwise, they are the result of acquisitions: as the majors are risk averse, small aggressive companies, frequently springing up from nowhere, venture into new high-risk plays; if successful, they then wait for the best moment to put themselves up for sale to the majors, or to some reserve-thirsty national oil company of an importing country, notably China. At the same time, the majors also tend to divest mature assets, which can more profitably be exploited by smaller companies with fewer overheads. As an increasing number of fields are aging and production is declining, the scope for this kind of specialization is bound to increase.

Producing countries and national oil companies: Different origins and mandates

The companies' landscape has completely changed as a consequence of these various tectonic movements, and today national oil companies (NOCs) dominate almost any ranking of the oil and gas industry, while international oil companies (IOCs) are struggling to keep their positions. However NOCs are a very diverse lot, and hardly constitute a meaningful category for analysis (Marcel 2006; Baker Institute 2007; Tordo 2011; Victor et al. 2012). A first fundamental distinction needs to be introduced between NOCs of net-exporting and net-importing countries. The latter frequently originate to exploit some limited national resources and then branch out internationally in the pursuit of additional resources needed to satisfy national demand. This is historically the case of many European NOCs (ENI, Elf, OMV, MOL) who were later privatized and have evolved in the direction of becoming more like IOCs; but it is also the case of the Chinese (CNPC, Sinopec, CNOOC) and Indian (ONGC, OIL) companies. Petrobras in Brazil was a similar case, though recent large discoveries at home have significantly reduced the company's drive to expand internationally.

Companies of net-exporting countries generally arise from the governments' perceived need to have a tool to be directly involved in the operations and management of the industry, whether in joint venture or on their own. Of course, it makes a huge difference whether the NOC is just a joint venture partner with IOCs or is expected to operate fields on its own: in the former case, it is likely that the NOC will have primarily a control function, and will not be forced to acquire the skills needed to operate autonomously. Further, among NOCs that are expected to operate autonomously, considerable difference may be found between companies whose operational and sometimes strategic autonomy is respected by the government, and are allowed to retain sufficient financial resources to invest and grow; and companies that are subjected to frequent interference on the part of political interests, drained of all cash from their respective governments and unable to invest as necessary (van den Linde 2000: 97–126). Among the former group we find companies such as Saudi ARAMCO and some that are projecting internationally such as Statoil of Norway (Gordon and Stenvoll 2007; Thurber and Istad 2012), Petronas of Malaysia (Lopez 2012) and PDVsa of Venezuela – until 2003 (Mares and Altamirano 2007; Hults 2012b). Notable examples of the latter are Pemex of Mexico (Stojanovski 2012), PDVsa since 2003, and Sonatrach of Algeria (Aissaoui 2001; Entelis 2012).

There appears to be some correlation between the nature of the state and of the political order and the kind of NOC that a country generates/allows. Countries aiming at short-run revenue maximization are likely to opt for allowing IOCs and confining the NOC to a supervisory role; developmental regimes that aim at economic diversification and industrialization are more likely to want their NOCs to be a tool to acquire know-how that may go even beyond the oil industry per se, and actively promote industrialization (through procurement policies and/or integration downstream in the value chain); finally countries where populist or clientelist forces are in power will possibly not resist the temptation to use the NOC as a cow to be milked for patronage or social endeavours that are not related to the oil industry at all.

The dialectic of national oil companies and governments and its impact on IOCs

NOCs are the creation of their respective states, but acquire a life of their own and engage in a dialectical relationship with their own governments, sometimes leading to a de facto reversal of positions, whereby the company becomes so powerful that it controls the government rather than the other way around. Recent history records at least two clear cases of companies that became autonomous and so powerful that they aimed at controlling their respective governments: Gazprom in Russia and PDVsa in Venezuela.

Gazprom (Stern 2005; Victor and Sayfer 2012) is the product of the transformation of the Soviet Ministry of Gas into an integrated company controlling all major gas fields and – crucially – the entire domestic and export pipeline network of the Russian Federation. In contrast, the dissolution of the Ministry of Petroleum gave birth to several independent companies, some of which were privatized in notoriously non-transparent fashion during the Yeltsin administration. The state, however, retained control of Transneft, the company controlling all oil pipelines. The bottom line was that the Russian state initially lost control of the oil sector, but the gas sector remained closely associated with the state. Indeed, being the main source of the country's foreign currency and the state's revenue, Gazprom became the salient power structure, for a period obfuscating not just the elected president but also the armed forces and the secret services. Thus Viktor Chernomyrdin, the former Minister of Gas (1985–9), then founder and first CEO of Gazprom (1989–92), became prime minister and was at the head of the government from 1992 to 1998. His successor as CEO of Gazprom, Rem Viakhirev, proceeded to partially privatize the company and obtained the right to vote for the shares that remained in the state's ownership – effectively turning Gazprom into a management-controlled entity.

When Vladimir Putin was elected president in 2000 it took him more than a year before he succeeded in recovering to the government the right to vote for the shares owned by the state, and then getting rid of Viakhirev and appointing in his place his own man, Alexei Miller. This marked the restoration of full government control over Gazprom after close to a decade during which it was rather Gazprom that controlled the government. It did not mark the end of the importance of Gazprom as a tool for power, domestically as well as internationally. It is not without significance that Dmitry Medvedev, who succeeded Putin as president and kept him as his prime minister, had previously been the chairman of the board of Gazprom.

The case of PDVsa saw the company becoming increasingly independent throughout the 1990s, pursuing a strategy of transforming itself into an IOC through downstream integration (acquisition of refineries in the Caribbean and of Citgo in the US), opening the door to foreign companies to invest in Venezuela and ignoring OPEC quotas. The wisdom of PDVsa's strategy has been hotly debated including by internal critics (Bouè 1993 and 2004; Mommer 1998; Mares and Altamirano 2007; Hults 2012b), but what interests us is the open clash with political authorities that followed the election to the presidency of Hugo Chàvez in 1999. Chàvez had attacked the company throughout his electoral campaign, and the company's CEO, Luis Giusti, immediately resigned. Chàvez appointed a succession of presidents to the company, with decreasing corporate legitimacy and acceptance, until at the beginning of 2003 the company revolted against the government and proclaimed a petroleum strike, which failed to force Chàvez out of office. Instead, the government re-established control over the company and fired some 18–19,000 of its employees, 40 per cent of the total workforce. The company has had a hard time recovering from such a huge haemorrhage of human capital.

These may be said to be extreme cases. More frequently, the company has de facto monopoly of technical information, and can influence government decisions through the way in which information is presented or selected. Governments relying exclusively on a national oil company put themselves in a position not very different from what they had when they faced a single foreign concessionaire – except that the national oil company is expected to pursue the national interest. However, in most cases what exactly is the national interest is a matter on which opinions may differ. Thus governments need to find a difficult balance between relying on IOCs and limiting the role of the NOCs to that of a controller (in which case the NOC frequently ends up being 'captive' of the IOCs); or nurturing a truly competent and autonomous NOC, which then may be difficult to control. Efficiency and competitiveness would dictate that the NOC be allowed to be as autonomous as possible, but then exercising control will become a technical and political challenge.

The blurring boundaries of oil and gas companies

The neat categorization of energy companies in national and international, or majors and independents is challenged by the multiplication of actors that are relevant to the functioning of the industry and the blurring of boundaries between different typologies of companies. Today, we have electricity or petrochemical companies integrating upstream and becoming producers of oil and gas as well as companies integrating downstream in the opposite direction. Oil and gas companies increasingly see themselves as 'energy' companies, monitoring and investing in a much wider array of energy sources.

I will here focus on two specific boundaries that have become increasingly permeable over the years: the boundary between hydrocarbons and finance and the boundary between industry and services. The relationship between the oil and gas industry and finance has become closer and closer. I mentioned already the fact that the opinion of financial intermediaries has become increasingly important in the definition of oil companies' strategy – something that was unheard of in the past, considering that the industry has traditionally relied on internally generated funds to finance investment, and on very low leveraging of equity. I also already mentioned the fact that it has frequently been cheaper to acquire reserves on the stock market rather than through the drill bit. More broadly, we may say that hydrocarbons have increasingly become financial assets, and are viewed and treated as such.

It is T. Boone Pickens, Texas wildcatter, corporate raider and financier, who formulated the theory that oil companies' worth is essentially the discounted value of their reserves viewed as a wasting asset and launched bold takeover attempts whenever companies' capitalization was below the intrinsic value of their mineral assets.[6] This view has become dominant, and essentially reduces oil and gas deposits to expected streams of future revenue. The development of futures markets has supported and reinforced this tendency, making it possible on the one hand to create 'synthetic wells' through the purchase of an array of future paper contracts, and on the other to anticipate monetization of physical assets through the sale of an array of future contracts. Oil trading has ceased to be the prerogative of physical traders and has attracted a growing number of banks, institutional investors and hedge funds. Today, you can buy oil from a bank and treat it as any other financial asset.

As oil companies feel the heat of competition in the financial marketplace, they have pared down their industrial assets (for a time before it collapsed, Enron was considered a model company because it had so few physical assets) and increasingly become intermediaries. The supermajors advertise themselves on the basis of their financial capability, that is, the fact that they can mobilize large funds either internally or through debt – preferably off their balance sheet, that is, through project financing. Thus the financial function within the company has become more important, together with the legal, and the ability to coordinate large number of outside suppliers.

The tendency towards outsourcing – itself a consequence of concern for cutting costs and increasing profitability, under the stimulus from equity markets[7] – greatly contributed to the rapid growth in service companies. Today, almost all activities that are needed for discovering, producing, refining, transporting and distributing oil are carried out by independent providers or service companies. Increasingly it is service companies that are at the frontier of technology and know-how, and make their expertise available to whoever is able to pay for it. This means that there is precious little truly proprietary knowledge left in the hands of the largest oil and gas companies, and competitors can have access to almost all that they may want through appropriate service providers.

In turn some service providers are increasingly attracted into being operators as well. The line becomes very thin when, for example, providers of oilfield equipment and services may be requested to also provide actual oilfield management; and their remuneration may be based not on a fixed fee but on a success fee that will see their return increase if they achieve a higher output; finally, they may be offered remuneration in kind (oil) rather than cash, or be requested to take a small equity stake in the venture in order to reduce the burden on the main promoter and enhance the incentive to achieve optimal results. At the end of the day, the service provider becomes an oil producer. Blurring of the lines along different stages of the value chain also occurs with trading. Thus, producing companies establish trading subsidiaries, but traders also increasingly invest in physical assets such as deposits, refineries and in some cases also producing assets (as is commonly the case for other commodities, notably agricultural or metals).

How exactly this is going to influence the dynamics of corporations and states is not altogether clear, although it appears that it may favour the position of states. On the one hand, it is certain that the rise of the service companies greatly supports the ability of many national oil companies to achieve levels of proficiency similar or equal to those of the largest international oil companies, hence greatly reducing the contractual power of the latter. On the other, one may say that control of technology and all forms of know-how remains in the hands of foreign – essentially OECD-based – providers. Yet in the end it is control of the commodity – oil or gas – that allows appropriation of the bulk of the rent: service providers, even as they evolve towards being producers and even if they might be very profitable companies, in the end operate in a competitive environment in which rents are difficult to appropriate. As long as physical production remains primarily in the hands of state-owned companies, it is the states that will enjoy the rent.

Shifting oil trade flows and the rise of state capitalism

The processes highlighted so far – the growing role of national oil companies, the tightening financial pressure on the international majors, the blurring line between oil and gas companies, electricity producers, service companies, traders and financial institutions – must be viewed in the context of shifting flows of international oil trade. In fact, at least three major trends promise to change the direction of major international oil flows, simultaneously affecting the role of companies:

- The rapidly growing demand for oil and gas in Eastern and Southern Asia, dominated by China and India (BP 2011; EIA 2011a; IEA 2011).
- The growth of oil and gas production from non-conventional sources, which promises to make North America essentially independent of net imports from the rest of the world (BP 2011; EIA 2011a; IEA 2011; Morse 2012).
- The adoption of more stringent decarbonization objectives, if not at the global level, certainly at least in the European Union (European Commission 2011a and 2011b).

The combination of these trends appears to open the possibility for a progressive regionalization, if not bilateralization, of the oil market; a process of segmentation in separate submarkets, within which separate regimes would prevail and between which arbitrage would be less than perfect.

The rapidly expanding markets in China, India and the other leading Asian countries are closely controlled by the state through their respective NOCs and a variety of administrative tools. The NOCs of the Asian importers are projecting internationally in order to find new resources to satisfy their growing domestic demands. They have been aggressively investing in all continents – including in North America – and seek a blueprint for cooperative relations with the NOCs of the Middle East. It is in fact clear that net additional supplies available to them will come primarily from the Gulf. Success on both counts is not guaranteed: notwithstanding the considerable visibility of and alarm raised by Chinese investment in the upstream in Africa, the Middle East, Latin America and even North America, outcomes in terms of additional production have not been stellar. As for cooperation with the Middle East NOCs, the Chinese, Indian, Korean, Malaysian companies have received no preferential treatment for access to upstream resources. Rather, there is an emerging pattern of cross investment in refining and petrochemicals: Kuwait's KPC and Saudi Arabia's Saudi ARAMCO and SABIC are investing in China, while Sinopec is investing in Saudi Arabia.

But the most remarkable feature of the emerging cooperative relations between the Gulf countries on one hand and India and China on the other is that they are being sanctioned in a context of greatly enhanced bilateral relations at the highest political level. Top leaders of both sides pay highly visible state visits – while in the past relations were tepid or non existent. Political relations are the framework within which reciprocal guarantees concerning supply and demand are exchanged. In contrast, the OECD countries constantly request security of supply from the Gulf exporters, but are either unwilling or unable to offer the security of demand that the exporters expect in exchange. Thus distinctly state capitalist relations prevail in the growing interrelationship between the Middle East and Eastern and Central Asia, eventually leading to some segmentation of bilateral or regional trade from the rest of the world.

In North America, a very large number of producers compete in a highly liquid and sophisticated market, which is as close as it might be to a competitive ideal. This market will remain open to imports from the rest of the world if needed; but as prices in North America are today lower for both oil and, especially, gas than prices elsewhere in the world, the question of potential exports from North America is raised. A debate is therefore open in the US congress and between the US and Canadian governments concerning the extent to which export projects should be allowed or encouraged – or should North American hydrocarbons not be reserved to continental use, notably to support the revival of American industry? The possibility that the door to exports might remain only partially open, consolidating a price differential between North America and the rest of the world is therefore very much in the cards.

Finally, the quest for decarbonization is pointing to a decline in the demand for oil in the OECD countries, which are the main market of the IOCs. In the case of Europe, the future of gas demand is also surrounded by considerable uncertainty, which reflects especially negatively in relations between the European Union and Russia. There two completely opposite approaches are confronting each other: Russia's is pure state capitalistic, maintaining the export monopoly of Gazprom, insisting on indexation of gas prices to oil prices, and demanding long-term demand security; while the EU's is completely focused on establishing an integrated and competitive internal market and on reducing GHG emissions.

Thus the IOCs can look forward to being leading actors in North America, yet be subject to significant competitive pressure and continue serving a shrinking European market, but with respect to the growing flows elsewhere in the world their role is at best one of intermediaries (between NOCs of the producing and of the importing countries) – that is, traders, or technology or finance providers (competing with service companies or financial intermediaries).

Conflict or Cooperation?

Will then conflict or cooperation prevail in future oil and gas trade? A new global regime would need to be arrived at through interstate negotiations and an agreement that would necessarily encompass climate change as well as all energy sources besides hydrocarbons. We are very far from that – indeed even within the EU we have an agreement on decarbonization but complete polarization when it comes to preferred energy sources.

Thus it is more likely that we shall see the emergence of partial, regional regimes: distinctly state capitalist between the emerging countries of Asia and the Middle East, but probably also Africa and Latin America; market based and competitive but with de facto limitations to export in North America; while the European market will remain open and competitive but demand will be destroyed in the name of decarbonization. State capitalist relations do not guarantee absence of conflicts, but may succeed in allowing for expansion of flows and win-win relationships. North America will be less interested in global supplies and less prone to intervene to guarantee its own interests. Finally, Europe will be locked in a negative spiral of low-level conflict with her suppliers, high prices and declining domestic demand. IOCs are likely to be under attack from both sides – from the producers because NOCs will become more empowered and from Europe because they are peddlers of 'dirty' energy – and will see their role progressively eroded.

Conclusion

The oil and gas industry may not be typical of conditions prevailing for other commodities or indeed for the manufacturing industry, but oil and gas continue to dominate international trade – and it is clear that states always have been busy intervening in oil and gas affairs. Although past regimes, notably the regime of the sisters which allowed for the golden age of oil, were primarily defined by agreements among companies, states were never too far from the scene. In the 1980s and 1990s, states retreated and it was expected that the oil and gas industry would become entirely 'market based'. However, the influence of states never quite disappeared.

Since the turn of the twenty-first century, national oil companies have shown that they can effectively compete with major international oil companies, and have become more powerful. The system of international oil companies, various service providers, financial intermediaries and self-regulated markets has failed to provide an answer to emerging issues. We are not in a second golden age of hydrocarbons, far from it. And this is not because the resource base is becoming exhausted – as recent trends in non-conventional oil and gas production have made abundantly clear. It is rather because of excessive price volatility and difficulty of precipitating enough investment to meet future demand. As a consequence, pure reliance on market mechanisms is not sufficient guarantee that existing resources will actually be brought and made available to the market.

In the impossibility of reaching a global compact covering climate change as well as all energy sources, trade in oil and gas will see the coexistence of multiple regimes with separate, yet sometimes overlapping domains. While state capitalism will prevail in the most dynamic sector – increased exports from the Middle East to East and South Asia – market relations and competition will prevail in an increasingly independent North America. Conflicts are likely, especially between the European Union and her suppliers – primarily Russia and North Africa – due to the conflict of desired regimes and the aggressive decarbonization path embraced by the Old Continent.

Notes

1. The expression 'seven sisters' was originally coined by Enrico Mattei, the founding CEO of the Italian national oil company, ENI (Sampson 1975).
2. It did so through the Gulf Basing Point System as explained by Stevens in Chapter 1.
3. For a detailed and more precise account see Anderson (1981: 179–97). The US government supported the idea that oil revenue be divided on a 50/50 basis. The fiscal implications only became evident at a later date and were the normal implementation of existing legislation.
4. In 1974 the Subcommittee on Multinational Corporations of the Committee on Foreign Relations of the US Senate published a collection of previously secret documents under the title 'The International Petroleum Cartel, the Iranian Consortium and U.S. National Security', detailing the workings of the cartel and the role of the Iranian Consortium.
5. It was also a climate of intense nationalism, as underlined by Stevens in Chapter 1. Yet one should not forget that of the smaller Gulf countries Kuwait only became fully independent in 1961, while Bahrain, Qatar and the UAE followed in 1971. Thus the full political independence of these very important producers (from 1955 to 1964 Kuwait was producing more oil than Saudi Arabia) came very late, just before the regime change.
6. Pickens's Mesa Corporation launched a hostile takeover bid for Gulf Oil in October 1983 arguing that the company was destroying value because of the incompetence of its management, and the underlying value of the assets – that is, primarily the oil and gas reserves – was higher than the value of the company. He did not succeed – Gulf was taken over by Chevron that made a higher offer – but Mesa and all other shareholders gained big. He went on to attack Phillips and Unocal – again failing to take over the companies but forcing them into steps that netted Mesa major profits. Phillips was taken over by Conoco, Unocal eventually by Chevron, after a higher bid by CNOOC was refused.
7. It is very noteworthy that this tendency is found only among oil companies that are publicly traded in OECD equity markets – because they need to worry about the opinion of financial analysts. National oil companies are radically different: the opposite extreme is offered by the Chinese companies, which have upward of a million employees each, and internalize everything – including food catering or nurseries for their employees. Most National Oil Companies are far less prone to outsource.

7
Communities: A Case Study of Oil Sands in Nigeria

Murtala Chindo and Michael Bradshaw

The purpose of this chapter is to examine the key issues relating to the impact of mineral industries on communities and to demonstrate how the different stages of the extractive cycle come with varying scales of impact on communities. For a focused view on these issues, a case study of proposed oil sands development in Nigeria is presented.

Oil sands are of particular relevance given their growing importance in relation to global energy security and their consequences for climate change (see Criqui in Chapter 9). Although oil sands development is being pursued by the petroleum industry, the techniques of extraction are more akin to mining. Many African and Latin American countries are now witnessing the beginnings of oil sands operations in their regions. These operations represent an invasive form of mining that covers an extensive area of land, which exposes communities to both direct and indirect impact, as well as immediate and long-term negative consequences. It also raises wider issues about the costs of supporting a high carbon-intensive operation and export-driven energy investments in remote communities that are also ecologically vulnerable areas. The existing problems between communities and the oil companies operating in the Niger Delta region of Nigeria are well known, and a year does not pass without conflict in this oil-producing region. The recurrent conflicts result from years of disregarding the host communities' right to a clean environment and development. Nigeria is now, in an effort to diversify its economy away from conventional oil and gas, ironically moving towards developing another finite resource and a dirtier form of crude oil. If measures are not taken at the early stages of development of the oil sands, the result will only replicate or even exceed the tragedy of the Niger Delta situation.

The chapter starts by conceptualizing mining communities, and examines how the meaning changes depending on the type of mining and the scale of impact. The second section assesses the impacts of mineral development on mining communities by examining the cycle of development, starting with permits, feasibility studies, construction, production, and on to closure and reclamation. The third section turns to the case study of Nigeria and takes a broader look at Nigeria's energy wealth, asking whether communities have benefitted from a share of the oil wealth. The fourth section outlines the growing importance of oil sands for global energy security, and considers how Nigeria plans to develop this form of crude oil. The fifth section presents a case study of potential oil sands host communities, and demonstrates how identifying and consulting with the affected communities at the initial stage of resource extraction is not only best practice, but can minimize the eventual negative impact and reduce the chances of conflict. The chapter ends by drawing lessons on how to manage communities as an integral part of the mining cycle, regardless of changing ownership structure or regime, whether 'liberal capitalist' or 'state capitalist'.

Conceptualizing mining communities

Mining is frequently a localized activity, situated within rural and remote areas, where people live and make use of land for agriculture and other means of livelihood and where the forest is not tampered with, and biodiversity is not disturbed. Mining is also associated with the history and development of many local communities around the world. There are a number of examples where communities have grown out of mining and are good examples of what mining boom towns are like. Prominent examples of these include Antofagasta in northern Chile, Ballarat and Bendigo in Western Australia, Bannack, Montana (Rocky Mountain) and South Pass City (Wyoming) in the United States of America, Johannesburg in South Africa, Kiruna in Sweden, the mining towns in Lancashire and Northumberland in England and Jos in Nigeria. Mining in such communities has significant socio-economic and environmental impacts on the local population, including those living in proximity to mining sites, and those hosting migrant workers.

The whole notion of community in the social sciences is complex and highly contested (Helen et al. 2002; Kapelus 2002; Jenkins 2004); as such, this literature review does not focus on that debate. The terms 'neighbour', 'community', 'public', 'camp', 'village', 'host' and 'resource town' are variously used to refer to people, who host, are affected by, or have an interest in mining (Cheney et al. 2002: 5). For example, Veiga et al. (2001) define mining communities as the population in any area that is affected by a nearby mining operation. The definition and classification of mining communities is more fully developed by the International Institute for Environment and Development (IIED) in its 2002 report, titled *Breaking New Ground*. The Best Practices statement of the IIED divides mining communities into occupational, residential and indigenous communities. Occupational communities comprise households that depend on earnings from the mine and, as a consequence, have to seek alternative survival strategies once the mine closes. Residential communities either emerge during the operation of the mine or are in existence well before the mine's development. This group, both in close proximity or farther away, hosts mine workers and mine facilities, and is affected by the activities of the mine. Environmental mismanagement can directly and easily affect these communities, such as through water pollution and dust. Indigenous communities are local people, who have special attachment to their ancient lands and a cultural affinity to their territories prior to the discovery of resources in their locality (MMSD 2002: 200).

These definitions and categorizations are not mutually exclusive and change over time depending on the nature of the mineral and the characteristics and stage of mining operations. The IIED classification does not take into account communities that are potential hosts to extractive activities and whose lives have been affected by the mere presence of economic resources awaiting exploitation, that is, the impact of the possibility of mining operations. Nonetheless, the IIED's recommendations opened up the debate over classifying and understanding the nature of the communities, which are affected by resource extraction at any stage of development.

The geographic location of host communities relative to the mining area is significant in determining the scale of impact (Department of Energy and Resource Management 2011). The mining industry tends to define communities in geographic terms (AccountAbility and BSR 2004). While the definitions may vary according to the size of the company, they usually specify those communities lying within the impact zone of the operation. Depending on the situation, a community can also include landowners at the periphery, linkages with those downstream, and all those within the scope of a company's community development programme. For example, the International Finance Corporation (IFC) Environment Division (2000) sets out how a mining industry player in Namibia chose to define the whole country as its 'community' to fulfil its obligations to the whole country, rather than just a few neighbouring communities. In a different manner, a mining company in South Africa defines communities within a 50 km radius of its operations. This delineation based on geographic terms enabled the company to work with the communities in the vicinity of its operating licence. Luning (2011) argues that communities potentially facing damage from mining operations are often located further away. In order to illustrate this point, Luning (2011) cites Townsend and Townsend's (2004) case studies about communities that suffer from mine operations but are not classified as a mining community by a company. For example, the litigation case against BHP's Ok Tedi mine was brought by communities outside the impact zone but whose land was being affected by these operations. For any scale of mining development, there is the potential for communities in close proximity to be displaced and for local environments to be destroyed. However, 'mining communities in different geographic locations may differ widely in terms of culture, environmental characteristics, and collective attitudes towards mineral resource development' (Schafrik and Kazakidis 2011: 87).

The actual impact and the perceptions of the community will also depend on the role played by the government, the extent of its engagement with the community, the social changes brought about by mining and even pre-existing social conditions in the local community. As mining communities become more dependent on the extractive company, they concurrently become unable to diversify (O'Hagan and Cecil 2007). Diversification according to O'Hagan and Cecil (2007: 20) is the ability 'to distribute employment among different industries to reduce risk in the event of the demise of a particular economy'. Further, Bray (2003) and Giordano et al. (2005) suggests that a fair distribution of revenues earned from mineral rents not only provides opportunities for host communities to have infrastructure that would power development, but also reduces the possibility of conflict, whether upstream or downstream.

The impact of mineral extraction

Minerals are used to satisfy some basic (or otherwise) needs of humanity, which include the transfer of technology, increases in productivity, the generation of employment and skills, generation of income, and industrialization (Bridge 2004a; UNCTAD 2007; World Bank 2008). It is also true that mining impacts on the livelihoods and environments of those citizens of humanity at the receiving end (Fonseca 2004). Firstly, these impacts depend on the nature of the mining operation, the size of the project, type of extracted mineral, type of exploitation method, the project lifespan, the nature and sensitivity of the surrounding physical and social environment, and the effectiveness of planning, pollution prevention, mitigation and control techniques. Secondly, impact can vary significantly, depending on the management of the mines and the implementation of the legal and regulatory frameworks. Kemp (2003) adds that communities become more vulnerable when the government fails to adequately protect them. Thirdly, the life cycle of the mining project is another important determinant of its impact (Figure 7.1). The phases of the mining cycle can broadly be classified into prospecting and exploration, mine construction and development, exploitation and production, and closure and reclamation – all with varying degrees of environmental degradation, human impact and resource depletion. Before mineral resources are harnessed they have to pass through certain phases; each stage gives rise to particular impacts in terms of scale and intensity. According to Noronha (2001), the World Bank and the International Finance Corporation (2002), the impacts are inherently detrimental to the communities that are in close proximity to newly established mines or those that are closing down.

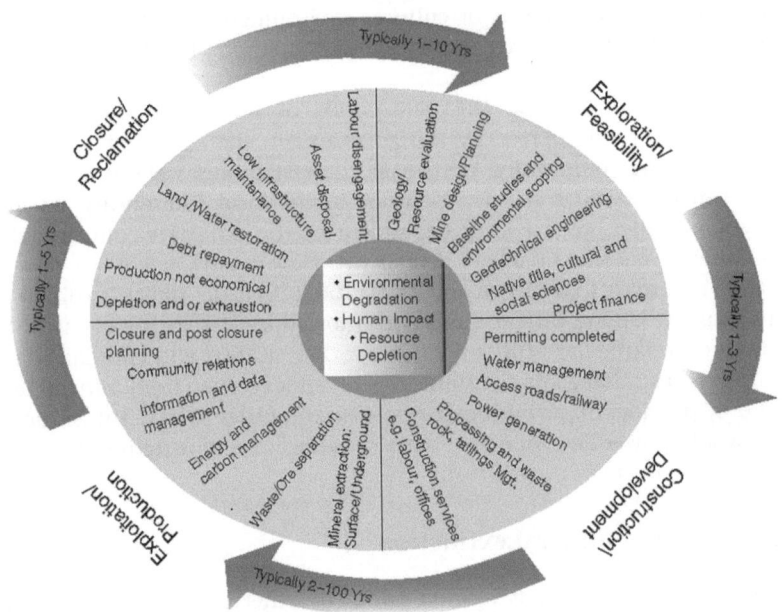

Figure 7.1 The stages of the mining cycle
Source: Author's graph.

During the initial exploration drilling, there is the risk of oil spills from oil rigs, which affect local ecosystems. This is among the first points of contact between the company and communities. This stage, however, has a low economic impact because it requires skilled manpower. The development of infrastructure involves the procurement and installation of materials and services. While production wells and refining plants are built, and pipeline corridors are established in oil production, for minerals, depending on the choice of extraction method, waste dumps and processing plants are built. This stage is characterized by the opening of remote areas, the construction of infrastructure and the stimulation of migration. This stage, though short, has long-term social and physical implications – it requires a large labour force, communities are displaced, and biodiversity is undermined.

During the production process, oil and gas are pumped from the reservoir through formation pressure, artificial lift and advanced recovery techniques until economically feasible reserves are depleted. The actual removal of minerals from the earth surface and sub-surface using either surface or underground technique forms the production stage in minerals extraction. Surface excavation can reach a depth of 460 m and more. The Palabora copper mine of South Africa has a width of about 2,000 m and depth of 762 m. The Bingham Canyon copper mine is up to 1207 m deep, making it one of world's largest open pit mines. At this stage, the impact of oil and mineral production includes leakages and spills, gas flaring and waste dumps, which furthermore contribute to environmental pollution, health problems, and security challenges. For example, Nigeria has a total network of about 5001 km of oil pipelines located within a land area estimated at 31,000 km^2 (Achebe et al. 2012). Pipeline leakages and fire outbreaks affect people, farmland, mangrove vegetation, aquatic life and water quality (Aroh et al. 2010).

Mine closure, or decommissioning in the case of oil, is the process of rehabilitating the land surface to what it was before. The oil wells are filled, tailings are disposed of, pits are protected, equipment is dismantled and removed from the site, and the environment is restored. The mining licence or title is then returned to the issuing authorities. At this stage, the impact from production could be made worse depending on the efficiency of the regulatory regime. Poor reclamation has a record of causing environmental pollution and land damage to surrounding areas. But also innovative mine restoration can benefit communities and the environment long after the mine has closed down. For example, the 'Mine Lake of New South Wales, Australia' is being used as a water hazard in a golf course (Evans 2006).

The following briefly reviews the literature with regard to the social, economic and environmental impact from mining, at the community levels. Social impacts are any mining-related changes in the way that people live, work and relate within a community. The impact can also cause changes in values, norms and belief systems as well as social organization. The impacts on the community are felt in the short or long term, depending on the cycle of development. Some of the costs that come with mining are summed up by Fonseca (2004: 23) as

> appropriation of the land belonging to the local communities, impacts on health, alteration of social relationships, destruction of forms of community subsistence and life, social disintegration, radical and abrupt changes in regional cultures, displacement of other present and/or future local economic activities.

In addition, social impact includes social ills – alcoholism, promiscuity, drug abuse, women and child abuse and the spread of HIV/AIDS among others, and an increase in violent crimes and idleness. Worse still, local communities experience the uncompensated loss of farmland.

The fear of the social risks associated with mining has become a major global concern of late. The industry is obliged to adopt recognized standards to minimize social risks and to adopt a proactive approach to social issues throughout all phases of operations. The MMSD (2002) was the first major industry-wide attempt at understanding the challenges faced by the industry (e.g. use, control, impact and management), and setting an agenda for change (e.g. partnerships, best practices, management of impact). While mining companies have adopted some of the recommendations for change, the report has had some criticisms (see Whitmore 2006). Cases of the failure to incorporate social impact assessments into mine plans and operational processes have been well documented, and so an ample body of literature exists on this topic. As social awareness increases, debates revolve about 'who', 'when' and 'under what conditions' mining should be allowed. In this regard, Bastida (2001) reported that the World Bank had recommended the shutdown of the Ok Tedi copper mine, because of tailings and waste rock disposal into river systems, but the Papua New Guinea government decided to allow the mine to continue in order to mitigate the devastating economic consequences of closure. The management of social issues in the mining industry now takes the form of community-engagement practices, where a 'social licence to operate' is obtained. This licence includes, among other criteria, acquiring free, prior and informed consent from potential host communities (Salim 2003). This process is included nowadays in the permit process to give communities the leverage to negotiate the conditions of resource extraction. A body of literature (e.g. Porter and Kramer 2006; Esteves 2008) has focused on the evolving guidelines which seek to drive companies towards engaging the communities as part of their social responsibilities. Doing so can reduce the potential for conflict and facilitate obtaining permits and accelerate development and production.

The level of economic benefits that communities derive from mining varies considerably from case to case. Rather than provide a catalogue of cases, this section presents the economic gains of mining to the local economy and how this has changed over time. Mining has the potential to create significant direct economic benefits in the generation of employment and skills, increasing income to the host country/region and the enhancement of domestic firms (Bridge 2004a; UNCTAD 2007). Mining can also contribute indirectly through linkages with the rest of the economy. Ritter (2000) considers employment creation and income generation as having the most important direct economic impact at the local level. The local community's ability to capture some or more of these benefits is dependent on (1) the scale of the mine and the maturity of the mineral project, and (2) the pre-existing community, its size, and the range of economic activities already undertaken (Ritter 2000: 5). The common examples of poor linkages with local productive structures are often linked to the lack of availability of local expertise or materials; for example, in the case of Ghana, major mining projects require specialized inputs, and so mining services are dominated by foreign firms because of local industries' inability to provide complex project services, such as steel work, electrical services, plumbing, ventilation repair and servicing of heavy-duty machines (Appiah-Adu 1999). Infrastructure that benefits local communities is planned and executed in the life cycle of the mine; once it closes, local authorities find it difficult to take on the responsibility of maintenance. Infrastructure can also be left under-utilized or allowed to dilapidate when a mine is closed (Dung-Gwom 2007). In this instance, Jackson (2002) recommends that infrastructural benefits should be negotiated prior to mining operations, so that the responsibility of maintenance after mine closure is defined and documented.

Mining can directly impact the environment through value chain activities – exploration and feasibility studies; construction and development; ore extraction, separation and dressing; refining; transportation; and closure and reclamation (Twerefou 2009). Until recently, the most common view on the environmental impact of mining has been that environmental impacts are localized as mines occupy a small area of land, when compared to the scale of other land uses, such as forestry and agriculture (Hodge 1995). However, a growing number of studies on the environmental impacts of mining challenges this view of mining as a localized activity with limited environmental impact (McAllister et al. 2001; Miranda et al. 2003; Bridge 2004a). The geology and chemical characteristics of the mineral extraction techniques and the size of the mine are critical to the level of environmental impact; hence, the larger the mine, the greater and more widespread the impacts. Proximity of the mine to habitation and economic activities, such as agriculture, fisheries and sources of water, also determines the extent of the impact (MMSD 2002; Miranda et al. 2003; UNCTAD 2007). Moreover, environmental consequences are not static but change in terms of scale and intensity from exploration to mine closure, often with long-term effects (Fonseca 2004). For Ritter (2000: 5), the 'remoteness' or the location of the mine relative to communities affected is of paramount importance as to the scale of its environmental impact. A study by Bridge (2004b) found that increases in investments in Peru, Chile and Indonesia intensified mining-related impact. The fear of negative environmental consequences often triggers rejection and political opposition by locals, who see themselves as the first to bear the negative impact (Bridge 2008b).

Global geographical shifts in the expansion of mineral projects to meet global demand and industry best practice have further intensified scrutiny of the impact of mining (whether local, regional or global) on biodiversity and ecosystem viability. The interplay between demand, supply and price significantly affects the closure phase of a mineral project. A prolonged slump in commodity price can result in the termination of production before the reclamation stage is reached. The sudden closure of a mine, either due to mineral depletion or when prices make it uneconomical to continue, and the resultant impact, can persist for centuries or even millennia (Sumi and Thomsen 2001). Mineral economics with weak institutions and environmental laws are even more vulnerable to the negative environmental impact caused by any sudden mine closure. Because of weak environmental regulations about sudden closure and poor monitoring in the late 1990s, the Namibian government and the host communities suffered from the sudden closure and withdrawal of foreign mining investors. The sudden collapse of tin prices marked the end of the economic opportunities it had created and significantly increased poverty levels in Jos, Nigeria. The importance of the industry to modern society underscores the need to promote improved environmental management in operations and to minimize the negative impacts. These factors therefore compel mining stakeholders to genuinely evaluate environmental issues at the community level in advance of mine development.

Nigerian oil and host communities

Nigeria is not only the most populous country in Africa but also the largest producer of oil on the continent. The political economy of Nigeria is based on extractive industries (Orogun 2010). Nigeria's rich oil reserves are found in the Niger Delta region, the Bight of Benin, the Bight of Bonny and the Gulf of Guinea. There are exploration activities in the deep and ultra-deep offshore and some activities in the inland basins of Chad, Benue, Sokoto and Bida. Nigeria produces 21.2 per cent of Africa's crude oil ahead of Egypt and Algeria. Nigeria also account for 2.6 per cent of global production (BP 2010: 8), and is ranked the world's eleventh largest producer of crude oil with a crude reserve of 37.2 billion barrels (Energy Information Administration (EIA) 2011a). Nigerian crude oil production averaged 2.6 million barrels per day (mbd) in 2012 and has generated over US$600 billion in the last 50 years. In 2010 alone, crude oil sales generated US$59 billion (Bala-Gbogbo 2011). Rather than bringing prosperity, especially to the communities where oil is extracted, oil has turned into a curse for Nigeria (Auty 1993; Karl 2007). The most comprehensive analyses of the oil industry and the Niger Delta communities are studies by the United Nations Environmental Programme (2011), Amnesty International (2009), United Nations Development Programme (2006), Watts (2004b, 2008) and Ross (2003).

Based on a review of the above literature, five decades of oil extraction in the Niger Delta have affected host communities in a number of ways. First, the petroleum governance system in Nigeria is weak and deeply flawed. Governments at all levels are often severely limited in their capacity, willingness or ability in enforcing environmental controls relating to oil and gas production and in delivering tangible and sustainable benefits to local communities (see Wagner and Armstrong 2010). Second, there is a near absence of benchmark studies on the impact of oil extraction on local communities compared to their health and occupations prior to the commencement of oil extraction 50 years ago. Third, communities were not consulted at the planning stages of oil development. Fourth, there is a lack of adequate measures to protect against human rights violation of the poor and deprived communities in the Niger Delta. Fifth, there is alleged collusion between the oil multinationals and government, which has fostered unhealthy interdependence between the two entities at the expense of the communities. For example, Iyayi (2006: 1) reported that

> local communities hold that it is the collusion between the oil companies and the Nigerian state not just to deprive them of their resources, but to set them against each other and exploit the resources in ways that have destroyed their environment and livelihoods that is the problem of the Niger-Delta.

Cawthra (2011: 46) further noted that 'the expansion of the activities of oil and gas multinationals in Nigeria, helped by the collusion of government officials, has played a direct role in fuelling corruption and violence and creating a seemingly paradoxical situation, characterized by the simultaneous enrichment of foreign companies and the impoverishment of local populations'. Finally, there are contested legal issues, such as resource ownership, revenue sharing, compensation and land ownership, between the local people and the federal government.

All the above issues culminated in community grievances and environmental degradation which led to violence (Müller 2010). The oil communities in Nigeria have suffered over time from lower economic growth and greater social unrest than the rest of the country. Poverty (using US$1/day as the standard) is endemic and the region is one of the worst examples of the resource curse, suffering from 'administrative neglect, crumbling social infrastructure and services, high unemployment, social deprivation, abject poverty, filth and squalor, and endemic conflict' (UNDP 2006: 9). Oil and gas operations for the past half a century have caused acid rain, respiratory diseases, damaged animal and plant life and contaminated the communities' air, land and water sources. The UNEP Report (2011) estimated that an initial capital of about US$1 billion is required to start what it calls the 'world's largest oil clean-up' in Ogoniland. The combination of poverty, deprivation, economic and political marginalization, poor governance, corruption and environmental degradation has spawned violent militia insurgencies with sabotage, looting, hostage-taking and ethnic separatism (Hassan et al. 2002; Lujala 2003; Watts 2004b: 278, 2008; Zalik 2004; Onyeukwu 2007; Orogun 2010). These challenges have not only affected the internal stability of Nigeria but have had an impact on the delivery of crude oil to the international market.

The lesson to take from the Niger Delta experience is simple – communities should be integrated from the initial conception of resource production to closure and even beyond. While oil sands extraction is potentially beneficial to the Nigerian economy, the issues raised above have so far received comparatively little or no attention in the solid minerals sector. Yet the level of impact of oil sands extraction could also cause displacement and changes to the local communities' environment, the people's wellbeing and their lifestyle. The planning stage of resource extraction provides opportunities for developing mutual understanding, consultation and dialogue about any potential impact of the project on the surrounding communities. Communities' involvement in that process is therefore of crucial importance. Through a range of techniques, the following case study uncovers the possible scale of impact of oil sands extraction on host communities in Nigeria in respect of the planning stage.

Oil sands: A new energy frontier

The abundance of oil sands reserves, along with other sources of energy, is being considered as part of the solution to meeting current and future global energy needs. This is a view held in common by the International Energy Agency (IEA) and the United States Energy Information Administration (EIA). Oil sands in their natural state are classified as non-conventional crude oil found in conventional oil reservoirs as heavy deposits of tar sands. Box 7.1 simplifies the misconceptions behind the interchangeable use of the terms 'oil sands', 'tar sands' and 'bituminous sands'. Large deposits of oil sands are found in a number of areas of the world. However, two countries in the Western Hemisphere contain approximately half of the world's total reserve; Canada and Venezuela account for more than 50 per cent of oil sands and 85 per cent of heavy oil (Meyer et al. 2007). Globally, natural bitumen is found in 586 deposits in 22 countries with major deposits in Canada, Russia, Kazakhstan, Venezuela and the US. Canada alone has about 227 natural bitumen deposits. Romania, Trinidad, Nigeria, Albania and Madagascar have significant resources.

Box 7.1 'Oil sands' and 'Bitumen'

'Tar sand' or 'oil sands' are composed of the aggregate mixture of heavy oil, mineral-rich clay, sand and water excluding any related natural gas. The heavy oil in tar sand is called bitumen. In its raw state, bitumen is simply a thick, black, sticky form of crude oil which is so heavy and viscous, that it requires heating or dilution with lighter hydrocarbons to make it liquid enough to transport by pipeline. The term 'tar sands' has been applied to bituminous sands since the late nineteenth century; however, from the mid-twentieth century experts have argued that the use of 'tar' to describe bitumen deposits is incorrect since tar is produced by the destructive distillation of coal and is chemically distinct from bitumen. While bitumen looks like tar, it can be naturally occurring or be a refined crude oil residue or can be rinsed and refined to produce fuel. The name 'oil sands' instead of 'tar sands' is now commonly used in Canada and other producing countries.

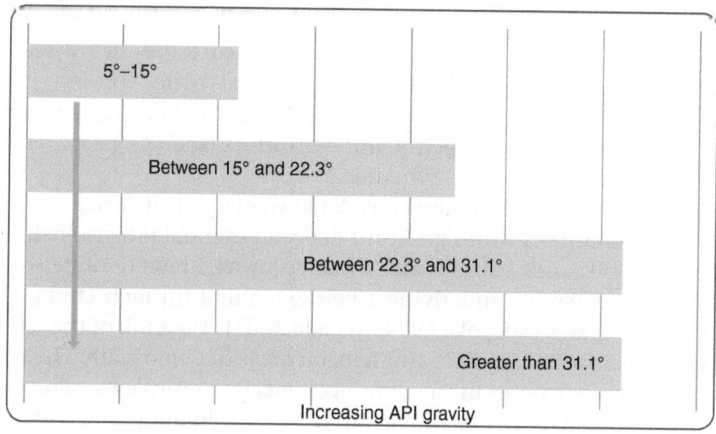

Figure 7.2 The fragmentation of oil based on API oil classification
Source: Author's graph.

Natural bitumen differs from light conventional oils by its viscosity at reservoir temperatures, its heavy metal content and its amounts of nitrogen, sulphur compounds and API gravity. In Figure 7.2, light crude oil has an API gravity greater than 31.1°, medium crude oil has an API gravity between 31.1° and 22.3°, heavy crude oil has an API gravity between 22.3° and 15°, and extra heavy crude oil/bitumen has an API gravity from 5° to 15° (Canadian Centre for Energy Information 2009). The bitumen extracted from Alberta's oil sands is generally less than 12° API, but varies by deposit. Nigerian bitumen has an API density of between 5.3° and 14.6°, while extra heavy oil is less than 10° and heavy oil is up to 20° API (Kazeem and Ademola 2010).

Generally, any crude that is below 15° is considered bitumen. In a number of cases, natural bitumen contains heavy oil by virtue of variations in the chemistry, depth and temperature of the reservoir. Regardless of these differences, high viscosity makes bitumen immobile under reservoir conditions. In Figure 7.2, the API gravity of light crude is greater than 30°; at this measurement it is lighter, it floats, and can easily be extracted from the reservoir to the surface by gravity. When bitumen API gravity falls below 10.0°, it becomes highly viscous so that it cannot be transported by pipeline to the surface without being heated. Bitumen can be derived as a by-product of petroleum, or can be upgraded to an API gravity of 31.1° (light crude).

Oil sands extraction differs from traditional petroleum recovery methods. Two approaches are common in extracting oil sands depending on the nature and depth of the reservoirs: (1) surface mining for subsequent extraction and upgrading of bitumen values, and (2) the in-situ method of reducing viscosity of oil sands so that it can flow to a producing well. For geotechnical reasons, the conventional method is most efficient on large-scale oil sands deposits of no more than 75 m of overburden. If the bituminous oil sand is buried too deep (from 75 to 400 m below sea level) for surface mining to be practicable or economical, extraction using in-situ operation becomes an option. Once the oil sands have been extracted, these can be converted into either petroleum derivatives (such as grease, wax and asphalt) or synthetic crude or 'syncrude'.

Nigerian oil sands reserves are considered the largest in Africa and in the global top 10 countries with significant deposit potential (Meyer et al. 2007). The Nigerian oil sands belt has an estimated proven reserve of between 31 and 42.7 billion barrels underlying a belt extending through Ondo, Ogun, Edo and Lagos states (Adegoke 1974; Ayoade 2007). The bulk of the oil sands are located mainly in Ode-Irele and Agbabu areas of Ondo state. The proven reserves of oil sands occur in the same ecologically sensitive mangrove environment of the Niger Delta, where current oil exploration and production is taking place. The deposits are concentrated along the upper fringe of the delta in equally sensitive agricultural environments (Adegoke and Ibe 1982). Documented reports indicate that the existence of oil sands in Nigeria dates back to about the early part of the twentieth century when the defunct Nigerian Bitumen Corporation (1907–14) carried out an appraisal on the suitability of oil sands for road surfacing. As a follow-up to this initial investigation, others drilled a number of boreholes in the outcrop zone with a view to determining the possibility of obtaining heavy oil through in-situ techniques. The most extensive exploration was conducted by the geological survey team from Obafemi Awolowo University in the period 1974–80. Later studies confirmed that the Nigerian oil sands are an important source of energy as well as an alternative source of hydrocarbons and raw material for the local petrochemical industries. The economic advantages that oil sands bring to producing countries have rejuvenated Nigeria's commitment to exploiting the resources for import substitution and additional revenue.

Case study: Potential oil sands host communities

The dominant ethnic group in the oil sands region is the Yorubas, but there are also Ilajes, Ijaws, Edos, Esans, Ikas and Afenmais ethnic groupings among the population. Farming, fishing, carving, lumbering, petty trading as well as blacksmith trades are the mainstays of a traditional economy and way of life. Farming is the major occupation, especially for men, but women also own and help in the farms and are engaged in the trade of excess farm produce; surplus farm produce is sold by women in exchange for household consumables. The remote communities are some of the most economically and politically marginalized populations around the oil sands belt. They have relied on the natural environment for survival in a way similar to the indigenous peoples of Canada. In the urban areas, people are engaged in manufacturing, administration, commerce, education, property development, banking and government, especially in the major cities. The space covered by the study area is that of a rural settlement pattern and the major communities that are covered by the study include the Ode-Irele, Omi and Ajagba. The rest are rural and remote communities, Legbogbo, Araromi, Gbeleju-Oke, Akingboju and Gboge. Basic infrastructure that can be found in the case study includes electricity, hospital, water, roads, schools and so on and these are located in big cities and local government council headquarters. Small settlements, especially villages, lack these amenities. For instance, Ijuba-Ijuoshun and Gboge are the most impoverished communities without any infrastructure provided by the government. The nearest health centre to Gboge community is about 17 km, while the closest primary school to Ijuba-Ijuoshun takes about an hour's walk.

The planned surface extraction of oil sands will be visible for the duration of the extraction project, affecting humans, animals and plants in the entire area of production and in adjacent communities. For example, 4750 km² of land is dedicated to surface mining of oil sands in the Athabasca Region (ERCB 2009). The extraction and production of Nigerian oil sands will follow a similar pattern. In 2006, the entire oil sands belt was delineated into six blocks, with a total area of 4103.91 km² (Ayoade 2007). These blocks of oil sands lie beneath the lands inhabited by these communities for centuries. Therefore, the communities lying within these blocks deserve key attention not only because they are located within or nearby, and will be impacted by developments, but because of their status under the 2007 Nigerian Mining Act, and the 1979 Land Use Act.

Numerous problems that will undoubtedly affect the individual, family and community are therefore envisaged, if an oil sands project comes into being. The views and perceptions collated through focus group discussions and interviews with community leaders, youths and women tend to be mutually reinforcing; the communities identified what they perceived were the positive and negative impacts of oil sands extraction. Loss of land for farming, fishing and small-scale forestry purposes, the fear of displacement, the rising costs of living and crime were generally considered negatively, while the provision of infrastructure for health and education, business opportunities, increased personal income and job creation were considered positively. Since the announcement of the commencement of oil sands operations in 2003, the communities have seen increasing settlement of people and commercial activities in the region, especially in the state and local administrative capitals, such as Akure and Irele. An increase in house rent, land value speculation, the use of illicit substances and alcohol, social fragmentation, environmental pollution and cultivation disruptions are some of the impacts felt by these communities at the early stage of oil sands development. The first visible impact of oil sands operations begin with the displacement of case-study communities (with an estimated population of over 100,000 people) from their traditional lands. This is because oil sands mining is so destructive to the territories and ecosystems that little in the way of traditional rural life is likely to survive in its vicinity.

The environmental impact of mining oil sands, such as greenhouse gas emissions, heavy metal contamination of water and soils, acid drainage, land occupation, tailings legacy and associated socio-economic risks have been of topical concern (e.g. Woynillowicz et al. 2005; Charpentier et al. 2009; Dogaru et al. 2009). Environmental issues present a great challenge to the development of oil sands in Nigeria. Information gathered across these communities reveals considerable and widespread anger about the environmental impact of oil sands development, which is primarily negative. This is because mineral communities in Nigeria have largely borne the brunt of extractive activities. The extent of the impact on communities has varied with the environment of remote commmunities being particularly adversely affected. Table 7.1 summarizes the problems associated with the environment that the communities expect, and how these are likely to affect them. As indicated in Table 7.1, the main environmental issues identified by local communities, such as loss of biodiversity, destruction of habitats and water quality, and pollution (gas emission, dust and noise), range from the physical presence of oil sands to those resulting from past and present preliminary drilling works done by government and companies. These perceptions are

Table 7.1 Perceptions of environmental impacts of oil sands

Concerns	Consequences
Water availability and quality	Pollution of fishing water, and impact on other aquatic organisms such as crabs Diversion of communal sources of water Long distance travel to access water because of pollution of proximal water sources by natural bitumen Flooding caused by barriers placed on natural-flow canals Contamination of surface and underground water making it unsafe for drinking and cooking Reliance on unclean water causes diarrhoea and dysentery
Loss of biodiversity and destruction of habitat	Cutting of forests and plantation at varying operational scales limits local opportunities for employment, food and income Reduction in the availability of arable land Loss of primary sources of protein and medicine Extinction of certain animal and plant species
Pollution-noise, dust and gas emissions	Pollution of the quality of air to the surrounding environment Cause of variation in rainfall pattern Contraction of airborne diseases such as those affecting the lungs Long-term hearing problems Machinery noise and movement upset domestic animals and children

backed up by what has happened in the oil and gas sector, where oil production is contributing to climate change and pollution that is hurting people, plants, animals, birds and fish (Watts 2008; Amnesty Report 2009; UNEP Report 2011).

Even as the capacity and the ability of the large number of local, state and federal government agencies and departments to address the above issues is in doubt, a rather optimistic view was also expressed:

> Both oil sands development and environmental protection can be tackled simultaneously. With sincere government commitment to complying with global environmental standards in oil sands operations; oil sands can be developed for the benefit of host communities and the Nigerian economy, while also protecting and meeting environmental sustainability.
>
> (Legbogbo Youth1, Pers. Comm., 2010)

The communities in the case study also fear that they are not likely to benefit from the royalties and taxes that will be paid by operators. Mineral taxes and royalties are paid directly to the federal government, and only a meagre amount, if any at all, comes back to the host communities in the form of physical infrastructure. The millions of dollars realized by the federal government from past bidding processes is yet to benefit any of the host communities. Given that a limited amount of this money will be invested for the development of host communities, they fear that, after the wealth and resources are taken away, they would be left with the remnants of a degraded environment similar to that of the tin mines in Jos, North-Central Nigeria. Communities expect royalties to be paid to them directly; they are unaware that royalties are paid directly to the government by the mining companies. In any case, the communities desire the transfer of a certain proportion of royalties to them for the improvement of quality of life and to offset any hardship in the event of mine closure.

The lack of governance structure in place to ensure that communities receive information about the status of the oil sands project is also infuriating. In the first instance, the government was blamed for lack of communication and insincerity in the bidding process. The belief that companies are being selected secretly without consultation was another reason to apportion blame. Communities believed that the government may have finally decided to abandon the project, and would gradually shut down the moribund oil sands offices in Akure and Ore. Indeed, the communities have been vindicated regarding the criticism that government has not been communicating or relating with the local communities about the current status of oil sands project:

> Honestly, the communities are not in reality aware of the progress made except through the media. Our strategy is to reach somewhere before

informing them about all the processes taken so that they can be convinced that this time the government is serious. Hold on... right now, the state is involved and I think they are aware of all the steps taken. It is not easy to reach the rural communities and that is why the states involved are first informed.

(MSMD official Pers. Comm. 2010)

This assertion emphasizes the need for the government to communicate with host communities through community leaders about the steps taken to ensure that the project is operated by credible investors. At the time of writing in 2012, oil sands investors had not yet been identified by MSMD, so the researcher could not have access to information on a company's policy on local communities.

During all the group discussions, participants indicate how much they expect from the oil sands operators. These needs and expectations are rooted in the current development needs of the remote towns and communities ranging from improvements in water supply, sanitation, local health facilities and other social indicators of human well-being and development. For various reasons attributed to the deliberate marginalization by the local council, it has been impossible for the Gboge and Ijuba-Ijuoshun communities to benefit from the provision of improved infrastructure. Therefore, remote communities in particular are greatly in need of development and are hoping to see how transformational and powerful natural resources can be to the communities. The oil sands project is seen as the opportunity that could provide a solution to what they see as government neglect. However, they expect the government to shoulder its responsibilities for providing municipal infrastructure while the companies complement government efforts. These findings complement the report by Bridge (2004a) and other organizational reports, which show how dialogue between extractive industries and communities of interest reduces the risks of access, antagonism to investment decisions, destruction to facilities and untimely closure of operations.

Lessons from the case study

The exploration and development of oil, gas and mineral resources can result in shared as well as potentially conflicting interests between industry, government and the indigenous people. This case study in Nigeria suggests that communities are not always sufficiently engaged in the consultations for the development of oil sands. This can lead to misunderstandings, misinterpretation and disagreement, and in some cases, violence, if there is a communication gap or when those involved fail to be transparent and communicate openly and honestly with one another. To minimize misperceptions from community members, a single collaborative working group with representatives from all those in the affected areas would be better placed to address these issues. Representation from various groups such as youth and old people organizations, councils of traditional rulers, women and local business, would help to ensure that a broader range of issues are considered. Such a development would boost communities' confidence as being integral partners in the development of resources in their locality.

Host-community views and perceptions of the environmental consequences of oil sands extraction should be included in the management and monitoring of the environment. Indeed, participants from the communities remain resolute about maintaining a harmonious relationship with the government and potential investors, despite the enormous challenges posed by oil sands extraction to their livelihoods. This amicable relationship offered by the communities may be the best approach, at least initially, for long-term harmonious working relationships. This will forestall resentment among the Niger Delta communities, the government and the industry, which has led to confrontation and violence in the past. This shows that dialogue between extractive industries and communities of interest reduces the risks of antagonism to investment decisions and violence against operational facilities. It also helps to secure access to these resources, to minimize delays for regulatory approvals, and to gain access to service industry companies and local labour.

One of the challenges in the Nigerian extractive sector is the increasing conflict between operators and host communities. Perceptions of socio-economic benefits and environmental risks of resource extraction are particularly sensitive in communities that are remote, isolated and highly limited in terms of economic opportunities. One of the first steps in understanding the risks is an all inclusive study of the communities that are to be affected by mine development. Advancing knowledge of this nature follows international best practices, which recommend that community issues (such as their interests, concerns, priorities and collaborations) are identified and understood, options are provided for resolution of any conflict or misunderstanding and a foundation is laid for appropriate choices to be made in the future. Perception studies can also provide baseline information on the opportunities for local service providers, employment opportunities and how to avoid damaging impact on culturally or ecologically significant areas. Lack of similar studies in the early stage of oil extraction has led to community protests, conflicts and even confrontations that have affected and disrupted the Nigerian economy in relation to the global oil market.

Conclusion

The case study presented in this chapter emphasized the need for, first, identifying the communities within the zone of impact, and second, undertaking early consultation with the communities, because they should be considered as an integral part of mining/mineral development. The role of the communities has gone through certain shifts and changes with the changing global political economy. In the 2000s, with growing concern for environmental sustainability, mining communities have increasingly the power to stop mineral projects at any stage of development. It seems likely that communities in the near future will be considered an essential part of the resource itself and a determining factor in the production of oil, gas and minerals. The case study of the Nigerian oil sands has shown that communities can help shape government and extractive industries actions so that they are more conscious to the impact of resource development. Understanding community characteristics and addressing real and perceived concerns, including by helping to resolve disputes before they escalate, has the potential to reduce the chances of intangible expectations, campaigns, blockades, protests, sabotage, kidnappings or legal suits, thereby also increasing government revenue and reducing the costs to the company that such actions can generate. Therefore, countries that are scrambling to develop their resources must learn to consult and cooperate with their communities, even more so in resource curse countries like Nigeria, which have a bad reputation for resource conflict and underdevelopment of producing communities. Considering the vast area containing the Nigerian oil sands deposits, and the fact that it is being evaluated for future development, it is an ideal case for identifying some of the communities that are potentially at risk from these new extractive activities. This is because potential environmental and socio-economic impact can be mitigated, and in some cases, avoided at the project planning and design phase, which reduces the chances of conflict, and fosters cooperation for the maximum benefit of government, company and communities.

8
Sector Legal Frameworks and Resource Property Rights

Evelyn Dietsche

This chapter offers a legal perspective on the overarching argument of this book, which is that the world is currently undergoing an economic and political transition with profound effects on the oil, gas and mineral sectors. This chapter addresses the implications of the potential transition from the era of liberal capitalism to a new era of state capitalism in relation to the legal frameworks that are currently applied in the oil, gas and minerals sectors. The core premise is that property rights lie at the heart of natural resources-related conflicts in many countries and regions across the world.

Property rights are the 'outcome' of political processes that take place largely within the borders of sovereign nation states. They are formalized by sector legal frameworks, which include constitutional provisions, resource sector legislation and the contracts and agreements signed between primary resource owners and resource developers. Reforms to resource sector legislation undertaken in the 1980s and 1990s have been a critical component for cooperation between investors and governments which have supported large-scale, international investments in the oil, gas and minerals sector. In the era of liberal capitalism, sector legal frameworks were viewed as an important 'input' to decisions underpinning the allocation of capital investments and prioritized the perceived needs of resource developers for stable property rights regimes. The dominant view was that resource developers would invest in those countries where they found resource sector legislation to be most favourable (IBRD/The World Bank 1996; Campbell 2004).

In recent years this thinking has changed and has become broader and more inclusive. The new era of state capitalism accords greater recognition to the fact that nation states are not only there to grant to rights to resource developers but also have obligations and responsibilities to their domestic constituencies in regard to these resources. Governments are faced with the challenge of balancing these obligations and responsibilities against the interests of the resource developers. This challenge generates tensions around sector legal frameworks and the property rights that they incorporate. Examples where sector legal frameworks have come under pressure include debates around (i) the strengthening of host countries' capacity to (re-)negotiate contracts and agreements, the 'fair sharing' of resource rents, resource revenue mobilization for development and resource governance more generally (Bräutigam et al. 2008; Benner et al. 2010; Gajigo et al. 2012); (ii) the development of international frameworks around the recognition and protection of non-state third parties, including indigenous peoples and other communities adhering to traditional forms of governance (Southalan et al. 2011); and (iii) the development of new norms around the duties of nation states to protect the human rights of their citizens (Cameron 2010; Drimmer 2010; Southalan 2011; United Nations 2011).

In the popular and sector policy literature, this change in thinking is often interpreted as a negative trend. For example, discussions around 'resource nationalism' often suggest that policy changes affecting the *status quo* carry some notion of illegality or that they are invariably motivated by unproductive rent-seeking and other negatively perceived political responses (Stevens 2008a; Kretzschmar et al. 2010). This chapter argues that these negative implications are not necessarily the case and that the underlying change in thinking demonstrates a growing recognition that ·resource property rights and associated sector legal frameworks are not just economic factors influencing the commercial decisions of resource developers. They are also politically contested. And in order for such legal frameworks to remain stable over time they need to meet the demands of political legitimacy, shaped by the relationship between governments and their key domestic constituencies. This requires that resource property rights be treated in a more holistic way because without some form of socio-political backing sector legal frameworks alone will not be able to sustain large-scale capital investments. At the very least, by ignoring this wider picture those investing in the oil, gas and minerals sectors expose host and home countries to the risk of undermining the institutional bases upon which socially responsible market economies are based (Wells and Ahmed 2007; North et al. 2009; Acemoglu and Robinson 2012).

The chapter is structured in three sections. The first section provides a high-level overview of the sector legal frameworks currently applied across the different regions of the world. It is assumed that not all readers will be familiar with these. The following section explains the conceptual basis of the reform efforts undertaken during the era of liberal capitalism and highlights some of its intellectual flaws. It provides a summary of the so-called liberal school of property rights that has shaped the thinking of the 1980s and 1990s and juxtaposes it against three specific concerns critics have raised. The final section draws out the insights gained from the critique of the liberal perspective and discusses the high-level implications for the future development of sector legal frameworks.

Legal frameworks in the oil, gas and mining sectors

The legal frameworks applied to the oil, gas and mining subsectors support a common set of objectives. These include (i) to define the primary owner of subsoil natural resources; (ii) to set out how exploration and production rights are to be granted and to whom; and (iii) to specify the conditions for the participation of private and state-owned enterprises in the exploration, development and production of these resources.

Despite these common objectives, sector legal frameworks differ considerably across countries as well as the three subsectors. This section provides a systematic overview of the historical trajectories that have shaped the nature of the relationship between primary resource owners and resource developers across countries and regions. It is important to recognize these differences because this helps us understand how the approach to reforming sector legislation associated with the era of liberal capitalism has been shaped by insights drawn from preceding interpretations of the historical development of sector legislation.

Sector legal frameworks can be disaggregated into four tiers, illustrated in Figure 8.1. The first tier comprises a country's constitution, which typically defines primary ownership of subsoil resources. Next follows sector legislation. This typically comes in the form of a mining law and/or a petroleum or hydrocarbon law. Sector legislation further defines who holds primary ownership and sets out the rules and procedures for granting resource developers access to resources. It establishes the administrative framework for granting resource property rights in the form of concessions, licences, leases, or whatever other term is used in respective jurisdictions. It also defines under which conditions exploration or extraction may take place, and it may set

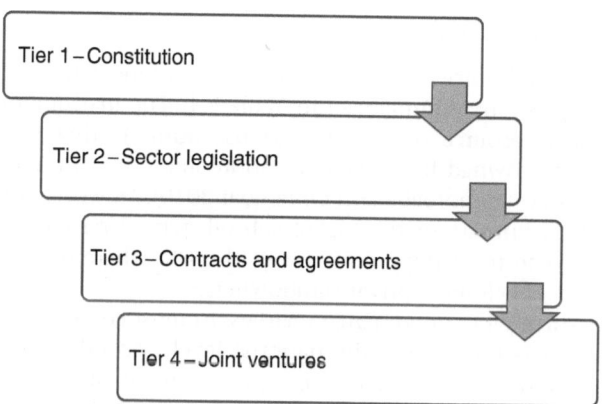

Figure 8.1 The four tiers of sector legal frameworks
Source: Author's graph.

out a fiscal regime. In some countries sector legislation tasks a national resource company (NRC) to look after the country's subsoil natural resources on behalf of the nation state. In these cases, it can be the case that NRCs grant exploration and production rights to resource developers.

The third tier comprises contracts and agreements, which define the relationship between the primary resource owner and the resource developers. Resource developers often consist of a consortium of two or more companies. Such consortia can include private extractive companies as well as NRCs, especially national oil companies (NOCs).[1] Contracts and agreements include licences and leases, concession and/or licence agreements, joint venture agreements between NRCs and private sector companies, production-sharing contracts/agreements, service agreements, various forms of mining agreements, and other country-specific types of contracts and agreements. Agreements and contracts also set out the applicable fiscal terms, where these have not been covered by sector legislation.

The fourth tier captures the contractual relationships between the consortium members comprising the resource developer, such as various forms of joint ventures (JVs), joint-operating agreements (JOAs) or similar. These relationships can include a NRC as a consortium member, and they can be incorporated or unincorporated.

The remainder of this section elaborates on each of the first three tiers of sector legal frameworks, but does not discuss the fourth tier in any greater detail. However, it is worth pointing out that where an NRC is party to the resource developer consortium established at the fourth tier, there is greater potential for conflicts of interests, because the same NRC may also represent the government in the signing of contracts and agreements. In such cases, the NRC finds itself on both sides of the negotiating table. This can have a negative impact on the clarity of its accountabilities.

Constitutional provisions

Cross-country differences in constitutional provisions can be categorized along three questions, summarized in Table 8.1. The first question is who is the primary resource owner? The global norm is that subsoil natural resources are owned by sovereign nation states and are referred to as domanial regimes (Omorogbe and Oniemoal 2010). However, a handful of countries grant primary ownership of subsoil natural resources to surface land owners. The most important case is the US, where on-shore subsoil natural resources belong to private landowners.[2]

There are three types of domanial regimes. In most countries the primary resource owner is the state at the national level. Second, there are countries where the primary resource owners are subnational state entities. For example, these may be regional states or provinces. Argentina is such a case. A third case comprises countries where there is no primary resource owner *per se*, but particular state entities have a reserved constitutional right to

Table 8.1 Differences in constitutional provisions

Tier 1-Constitution	*Who is the primary resource owner?*	Surface landowner (e.g. US onshore, some Canadian states, other legacy cases such as Trinidadian onshore blocs)
		Nation-state (e.g. Continental Europe and other countries with domanial regimes)
	What is the country's legal tradition?	Common Law (e.g. UK, USA, Australia, New Zealand, other former British Colonies)
		Civil law countries (e.g. Continental Europe) other legal traditons
	What are the origins and trajectories of sector legal frameworks?	Historically evolved (e.g. Continental Europe some part of Eurasia)
		Initially inherited from mother country in eighteenth/nineteenth/ early twentieth century, since then evolved
		Inherited from mother country in 1950s/1960s, fundamental reforms undertaken in 1980s/1990s

Source: Author's table.

grant administrative permissions that allocate ownership rights to private sector entities. For example, under the German federal constitution subsoil natural resources are defined as 'bergfrei' and neither belong to the freehold surface landowner nor to the state. Instead, administrative entities at the regional level can grant licences and concessions to resource developers.

The second question is what is the country's legal tradition? Legal traditions have shaped the process through which sector law has been developed. In civil law countries, where the law-making role is reserved for the legislature, statutes are the primary sources of law. Meanwhile, in common law countries the decisions of judges on cases brought before the courts play an important role in shaping sector legislation. For example, judges acting under common law played a critical role in the early development of US petroleum legislation. Their decisions set off a path dependency that shaped subsequent court decisions and enticed US private landowners to engage in competitive drilling, which in turn supported the rapid expansion of fossil fuels as a major source of energy in the US and elsewhere (Daintith 2010a, 2010b). When neighbouring landowners first started to enter into disputes over the exploitation of petroleum resources towards the late nineteenth century, common law judges applied the so-called rule of capture whereby the landowner who first lawfully drilled and pumped fluid resources from an underground reservoir could also capture the resources located under his/her neighbours' land. This rule had originated in the European context around the capture of wild animals, but had also been applied to water pumped from underground aquifers (Scott 2008). Acting under common law, American judges simply went for analogy and applied this rule to petroleum resources.

It did not take European governments long to learn quickly from the unintended negative consequences of the US experience. In the early decades of the twentieth century, they amended their legal frameworks to clarify that fluid subsoil natural resources also belonged to the nation state in the same ways that hard minerals already did (Daintith 2010b).[3] This sector legal change also affected the colonial territories that European countries were still controlling at the time, in particular the UK and France.

The third question is what are the origins and the reform trajectories of sector legal frameworks? While distinctions are not entirely clear-cut, there are broadly three groups of countries. The first group comprises mainly the more established European civil law and common law countries. The sector legal frameworks of these countries display path dependencies originating in their historic nation state formation processes, where ruling elites expanded and asserted monopoly rights over the use of power to control territory and people and in the process had to develop their public administrations (Bates 2001; North et al. 2009). Comparative economic historians point to resource-based regalia or tax farming systems as a cost-effective way for collecting revenue to sustain increasingly larger political and administrative entities that could gain and maintain control over territory and raise capital, not least to finance territorial wars (Tilly 1985, 1992; Levi 1989). Developing minerals provided ruling elites with revenue. But because there was little knowledge about existing deposits, sovereign rulers also needed to encourage miners to prospect and discover. This led to the development of the 'rule of discovery' and the principle of 'first come, first serve', whereby the miner who first staked a claim could also claim ownership over the resources discovered and produced (Bastida 2004; Scott 2008).

The second group comprises countries that gained independence in the nineteenth and the early twentieth century. This includes North American and Latin American countries, and countries in the Middle East and Eastern Europe which emerged from the break-up of the Ottoman and the Austrian-Hungarian Empire. Many of these countries inherited the legal traditions and sector legislation from their colonial mother countries. Subsequently, however, sector legislation has evolved along somewhat diverging paths shaped by each country's specific post-independence political and social history. While there are of course path dependencies, countries with similar backgrounds can also show considerable variations that evolved over the decades. For example, legislative frameworks now differ considerably across Latin America, despite these countries' common Hispanic civil law roots.

The third group comprises countries, which gained independence from the middle of the twentieth century and that typically inherited basic sector legal frameworks in the transition from colonial rule. Many of these countries share the experience of undertaking sector legislative reforms from the late 1980s onwards, influenced by the thinking that underpinned the era of liberal capitalism. Another feature that can be found among these countries is that the second and the third tier of the sector legal framework are not clearly distinguishable. In the absence of more fully developed sector legislative frameworks, contracts and agreements have been rubberstamped by parliamentary approval so as to lend them the same status as sector legislation. Some have argued that this has made it more difficult for countries to improve and develop more comprehensive sector legislation (Campbell 2004). Another serious issue affecting these countries is tensions and disputes around the formal exploration and production rights that have been granted to resource developers and informal land rights and land use practices usually associated with traditional forms of governance. The latter typically involve rural pastoral and subsistence farming communities, and/or internationally recognized indigenous peoples. However, this issue has also been a source of conflict in former colonies that gained independence before the 1950s and 1960s, for example in Latin America, Australia and Canada, where there are significant and/or recognized populations of indigenous peoples.

Sector legislation

Sector legislation further defines who holds primary ownership over subsoil resources. It specifies the authority that can grant exploration and production rights, and sets out more specifically the rights, obligations, limitations and conditions for maintaining and cancelling these rights. For example, sector legislation specifies if rights are to be granted by a sector ministry, an NOC or some other regulatory body. It also specifies the rules and procedures for accessing land, the use of other resources in the extraction process (e.g. water) and for developing infrastructure. Sector legislation sets out the general terms and conditions for the subordinate use of contracts and agreements and may further define the use, management and control over the rents derived from resource exploitation.

Variations across sector legislation can again be illustrated on the basis of three questions. These are summarized in Table 8.2. The first question is how general or specific is sector legislation? In general, oil and gas sector legislation is more general and less specific. A country's hydrocarbon or petroleum law may refer details to tier 3 of the sector legislative framework and support the use of standardized model contracts and agreements. Such models set out variables that can be altered, depending on the type and the conditions of the acreage in question. As a consequence, there is a lot of variation in oil and gas sector contracts and agreements. It is also quite common for a country to use various different types of contracts and agreements in parallel, each containing project-specific fiscal terms (UNCTAD 2007: Table IV.3).

Mining legislation is more diverse, and also more specific. This is because mineral resources include a wide range of high-value/low-volume minerals,

Table 8.2   Variations in sector legislation across the oil and gas and minerals sectors

Tier 2-Sector Legislation	*How general or specific is sector legislation?*	Oil and gas legislation is more general and less specific. It often makes use of standardized model Contracts and Agreements.
		Mining legislation is more diverse and more specific.
	How are resource property rights granted?	In the oil and gas sector rights are awarded in a number of different ways, including open-door systems and organized licensing rounds, based on administrative procedures or auctions.
		In minerals sector the most commonly used method is the *'first come, first serve principle'*.
	What role is assigned to NRCs?	Many OECD countries have privatized their NRCs. NOCs are particularly strong in the Middle East and they generally control the majority of the world's petroleum reserves. The strength of NRCs is more varied in other parts of the world.
		In the minerals sector, few NMCs still exist. They usually operate under special legal arrangements.

Source: Author's table.

such as diamonds and gold, metaliferious ores, non-ferrous metals, crude minerals and energy minerals. Often the mining law – which may also be referred to as a mining code, act or statues – sets out classifications for different types of minerals and associates different mineral regimes with these. For example, at times large-scale mining projects have been provided with specific sets of regulatory provisions and guarantees. Small-scale and artisanal mining is often governed by a specific set of statutes. The energy minerals, coal and uranium, are usually subject to very specific mineral regimes and contractual practices. And geographic areas where the presence of mineral resources is already well known may also be subject to special regimes. The mining law may also set out a variety of provisions in relation to taxes, customs, environmental standards and investment incentives.

The second question is how are resource property rights granted? Sector legislation usually sets out the regime for granting resource rights, through licences, concessions or leases, or whatever the term used in the respective jurisdiction. This is the first step in the process whereby the primary resource owner shares with, or transfers rights to, a resource developer or a consortium of resource developers. Rights are usually awarded for defined stages of the resource development process, including for example exploration licences, appraisal licences and production licences. These rights come with predefined limitations, such as the time period for which they are valid (duration), the specific work programme that has to be completed within this time period, the ability to sell or transfer them (transferability), the possibility to divide the licence (divisibility), as well as other conditions that affect the exclusivity, flexibility and the quality of title associated with the these rights (Scott 2008).

In the minerals sector, the most commonly used method for granting resource companies rights to explore and produce mineral resources is the historical principle of 'first come-first served'. It stipulates that resource developers are granted monopoly rights to explore and produce resources by the order in which their applications have been received. The origins of this principle can be traced as far back as the Roman Empire and to the practice of 'free mining' in medieval Europe, where it evolved against the background of providing incentives for prospecting and discovery when demand for minerals was rising but little was known about geological prospects. The principle has been strengthened in recent mining sector reforms in particular with respect to developing countries where there is still comparatively little known about the resources that may exist underground (Bastida 2006, 2008).

In the oil and gas sector, the systems granting resource property rights are usually not based on the principle of 'first come, first serve'. Licences can be awarded in a number of different ways and forms, sometimes used in parallel and decided upon on the basis of the knowledge the primary resource owner possesses about particular acreage and its associated geological conditions. There is a typical distinction between two types of oil and gas licensing systems (Cramton 2007; Tordo et al. 2010). Open-door systems award licences as a result of negotiations between the primary resource owner and the interested resource developer. Resource developers are either explicitly invited to express their interest in a specific geographical area, or can take their own initiative to approach the primary resource owner. Second, organized licensing rounds are based on administrative procedures where applications are evaluated against specific criteria set out upfront. Such criteria include, for example, the proposed work programme for exploration. Licensing rounds can be organized as auctions, where licences are awarded to the highest bidder. Different countries use more or less rigid mechanisms to evaluate the proposed work programmes and bids. Some countries determine upfront the fiscal and other elements that the proposals of applicants are expected to include. Others may leave all the parameters open. Countries may also vary these conditions depending on the type of acreage they offer.

Bidding is less common in the mining sector. It is typically used where geological prospects are well known, or they have been reasonably well established. The 'first come, first serve' principle is then replaced by some form of competitive bidding that puts the burden on potential resource developers to submit competitive assessments on the value of the respective geographical area. This practice is considered appropriate where national geological services have developed extensive minerals cadasters, or where national geological services or former state mining companies established the resource base. More recently, some mining sector initiatives have shown increased interest in enhancing geological knowledge to identify prospective areas that are suitable for the use of a bidding process (UNECA 2011).

The third question is what role is assigned to national resource companies (NRCs)? Some countries vest resource property rights in an NRC on behalf of the nation state. Where they are fully owned by the nation state, NRCs may be granted preferred or even exclusive access to licences and they may also be charged with the responsibility to oversee and manage the sector. NRCs will collaborate with private sector resource companies, if and when they require the latter's particular technological expertise and other business assets. Examples include most of the Middle Eastern NOCs, PDVSA of Venezuela and Pemex of Mexico.

The role of the NRCs is sometimes associated with ensuring that government can exercise direct control over the sector. However, technocrats dispute this claim. They maintain that similar levels of control can be achieved with both direct and indirect policy tools, where the latter involve applying different forms of state participation and regulatory oversight (Daniels 1995; Stevens 2003b; McPherson 2009). Over the past decades a number of OECD countries have massively scaled back the role of NRCs without generating the perception that their governments have lost control. Furthermore, NRCs have been associated with two challenges. First, they may be left to manage a rather complex and often conflicting set of interests and tasks. For example, in addition to acting as *de facto* primary resource owner setting the terms and conditions for granting exploration and production rights to resource developer companies, they may also be part of a resource developer consortium and thus be a business partner to the very same companies. Second, NRCs can add to the problem of domestic elites focusing on engaging in unproductive rent-seeking rather than re-investing profits from resource rents into the sector and the national economy more generally.

NRC involvement is particularly prevalent in the oil and gas sector (Victor et al. 2012). NOCs control the majority of the world's petroleum reserves, not least because of the particularly strong positions they hold in the Middle East (Marcel 2006). As pointed out by Stevens in Chapter 1, this is different from the situation that prevailed until the 1960s when the majority of petroleum reserves were still under the direct control of a small number of private companies. The power of NOCs is more varied across Latin America and Asia depending on the respective country's economic policy stance. NOCs also exist throughout sub-Saharan Africa, again with varying degrees of effective power, conditioned not least by the maturity of the respective countries' oil and gas sector.

NOCs have generally gained strength in countries outside the OECD, including in some that followed the liberal reform model in the early 1990s. For example, Kazakhstan has since revised its sector policies and re-established a NOC (Luong and Weinthal 2010). In other countries, for example Brazil, where Petrobras was partly privatized, the lines have remained blurred between the NOC operating like any other international private oil and gas companies and it being called upon as an instrument for meeting government objectives (de Oliveira 2012).

In contrast, in the minerals sector there exist only a few national mining companies (NMCs). Most former NMCs were privatized in the 1980s and early 1990s. Governments rely more on indirect instruments applied by the sector ministry and/or a regulatory body to control the sector. Those NMCs that still exist tend to operate under special legal arrangements, for example the joint venture agreements between DeBeers and Debswana and Namdeb, or the mineral arrangements in place with Gecamines in the DRC. Chile's Codelco and Sweden's LKAB are additional examples of the few surviving NMCs. But, some countries have recently re-established or strengthened existing NMCs, for example in Namibia, South Africa and Indonesia. This suggests that new types of sector legal arrangements might emerge in this context. An interesting observation is also that among the more success-ful mining countries are those that have retained NMCs, including Chile, Botswana and Namibia.

Contracts and agreements

There is an extensive literature on contracts and agreements, most of which refers to the oil and gas sector. This literature typically draws a distinc-tion between concessions and so-called contractual regimes (Johnston 2007; UNCTAD 2007; Tordo 2007). The legal difference between these two types of regimes rests with the exclusivity of the rights granted to private sector resource developers. Concessions assign rights to the licence holder (or 'con-cessionaire') when the resources are still in the ground. For example, the old-style concessions referred to by Stevens and Humphreys in Chapters 1 and 2 granted exclusive rights over vast geographical areas with long dura-tions, had few duties attached, and typically paid the primary resource owner only a small royalty. Modern concessions are much tougher and less exclu-sive, and their associated fiscal regimes usually not only include royalty but also various profit and value-based taxes, fees and charges.

Contractual regimes include both production-sharing contracts/agree-ments (PSCs/PSAs) and service agreement (SAs). The nation state – or the NRC as its agent – retains ownership over the reserves in the ground and the resources as these are produced. Rights to the resources produced are shared only once the resource developer has already been compensated for the costs incurred for bringing the resources out of the ground. Sharing happens either in the form of sharing production in-kind, or by paying private sector resource companies a premium fee for the extraction services they have rendered. Negotiating contracts revolve, among other issues, around the respective shares by which production is to be shared.

Contractual regimes are found mainly in the oil and gas sector, where they have been introduced from the 1970s onwards against the background of the negative experiences with early petroleum concessions. Over the past decades the use of contractual regimes has expanded considerably and they now dominate among the different types of petroleum fiscal regimes used. The more recent and emerging oil and gas countries tend to make use of PSCs, while Middle Eastern and other countries with strong and competent NOCs favour service agreements. More generally, in many countries outside the OECD it is quite usual to encounter a situation where, under the umbrella of generic petroleum legislation, various generations of concessions, production-sharing contracts and service agreements are used in parallel. Some argue that this situation might be negatively affecting the efficiency with which countries are able to govern the sector as a whole (UNCTAD 2007). Another concern is that stabilization clauses used in association with contractual regimes limit the ability of host countries to develop their sector legislation and apply regulatory and fiscal updates to existing contracts and agreements (Cotula 2008). In contrast, European and most other OECD countries generally prefer to use concessions. Concessions allow countries to evolve regulatory, fiscal and other relevant policy regimes, such as for example environmental legislation, without the need to change contracts and agreements. Governments are keen to retain the policy space and flexibility to expand or constrain the rights and duties of licence holders within the context of overall economic, financial and public policies.

Concessions continue to dominate in the minerals sector. PSCs are very rarely used, and in the few cases where such exist they involve low-value/high-volume and easily marketable minerals such as sand and gravel. The usual explanation for this is the greater diversity of mining products and the more specific marketability. However, some industry observers argue that the marketability argument no longer holds, because many minerals are now traded transparently on commodity exchanges. In general, there has not been much discussion or research into this issue. In those few cases where NMCs still exist, service agreements and joint venture agreements are usually applied. Table 8.3 summarizes the broad differences across the subsectors and across regions.

Alongside mining concessions, widespread use has also been made of so-called mining agreements, particularly in countries where more fully developed sector legislation was still missing in the 1980s and 1990s. These agreements have complemented sector legislation, and in some instances have been subjected to legislative ratification so as to lend them the same status as sector legislation. While the first so-called mineral development agreements emerged in the 1960s and 1970s with wide discretionary powers for government authorities, the trend from the late 1980s onwards has been for mining agreements to limit the scope and possibility for

Table 8.3 Sector and regional differences in the use of contracts and agreements

Tier 3- Contracts and Agreements	*Oil and gas sector*	Contractual regimes constitute the majority of petroleum fiscal regimes (i.e. production sharing contracts and service agreements). They are widely used by countries outside the OECD.
		Concessions are used by most OECD countries. Their advantage is that sector policy can be evolved without the need to change contracts and agreements.
	Minerals sector	Concessions are the norm. Particularly in Sub-Saharan Africa, mining agreements have been used in conjunction with concessions, often to fill investment-critical gaps in general sector legislation.
		Where they still exists. NMCs may use service agreements, or collaborate with private resource developers under joint venture agreements.
		PSCs are extremely rarely used.

Source: Author's table.

government discretion and to hold constant the parameters upon which resource developers have based their investment decisions. Modern mining agreements also refer to international arbitration to address potential investment disputes.

Modern mining agreements have been an important component of liberal thinking seeking to attract foreign investment into countries emerging from periods of conflict and economic crises. This rationale has been particularly relevant in the sub-Saharan African context, where mining agreements have supplemented sector legislation in the form of so-called investment agreements or investment promotion agreements (Bastida 2008). These agreements have often been explicitly designed to fill investment-critical gaps in general sector legislation. They set out the basis for determining the obligations of private sector resource developers, respective tax and other revenue payments as well as equity participation, and have often also offered extensive benefits and tax allowances to render the fiscal regime more investor friendly. They have catered to the needs of large-scale international mining investments during a period of time when the sector was suffering from a negative image concerning its profitability (see Chapter 2).

There are some differences in the use of mining agreements across Francophone and Anglophone sub-Saharan African countries. In Francophone countries, the so-called *convention minière* is commonly used to define the general legal framework applicable to a specific project in addition to the mining licence acquired under general mining law. For example, this is the case in Guinea, Mauritania, Niger, the Central African Republic and Senegal. The general mining law may include a model mining convention, on the basis of which a particular agreement is bilaterally negotiated between the government and the investor. Once agreed upon, an agreement may be subjected to ratification by the legislature. Agreements have typically also included extensive investment guarantees.

In contrast to the widespread use of mining agreements in sub-Saharan African countries, industrialized countries rely on unilaterally specified mining concessions to grant private sector mining companies access to mineral rights. These also specify the fiscal terms. The only exception is Western Australia, where large export-oriented projects and this state's heavy dependence on mining have required significant associated investments in infrastructure. In Latin America most countries also rely on concessions that are granted under national mining law.[4] Concessions are typically granted on the basis of a non-negotiated, non-discretionary procedure allocating ownership rights by way of applying the principle of 'first come, first serve'. Exceptions include cases where rights are held by former NMCs that have been privatized, or where known and studied deposits are still owned by NMCs. For other parts of the world there is less generic knowledge about mining regimes. In Russia, most Central Asian republics and other Asian countries, sector legislation and associated mining agreements and contracts are particularly varied and idiosyncratic.

The overview of sector legal frameworks presented in this section has sought to serve various purposes. It has set out in broad terms the historical trajectories that have shaped the relationship between primary resource owners and resource developers across countries and across the oil and gas and minerals sectors. Structural differences have been identified between the longer-established and more-consolidated countries, including mainly European countries and other members of the OECD, and other countries, in particular those that gained independence only in the middle of the twentieth century. With respect to the first set of countries, sector legal frameworks have evolved as part of the process of building and consolidating nation states. External factors may have triggered regime change, but internal factors have largely shaped the direction that change has taken. By contrast, in most other countries it has mostly been external factors that have provided the impetus shaping sector legal frameworks. First, there was the institutional heritage countries were left with at the time of independence. This was then followed by the political and economic ideas that have dominated in the respective post-independence eras. In particular, the liberal thinking that dominated from the 1980s onwards has focused on the specific objective of encouraging foreign investment. It has given little consideration to internal conditions, for example the perseverance of traditional forms of governance, or the developmental needs of host countries. The next section juxtaposes the thinking that underpinned liberal reforms with the concerns that critics have raised, and explores the importance of internal domestic factors.

Property rights: The liberal school and its critics

The objective of this section is to identify potential sources of conflict and collaboration in how sector legal frameworks have evolved, by drawing on valuable insights that the social science literature holds in this respect. The section is structured in two parts. The first part summarizes the liberal school of property rights, which shaped the ideas underpinning the sector legal reforms introduced from the 1980s onwards. The second part contrasts the approach taken to these reforms against three specific concerns over how the role of property rights and sector legal frameworks was interpreted in the period of liberal capitalism.

Property rights are a well-established analytical concept in the social science literature. As a concept it incorporates an intellectual discussion that is more specific than the summary of sector legal frameworks presented in the previous section. However, the concept has also been applied to investigate a wide range of economic, political and social challenges in relation to the governance of natural resources (Bates 1989; Alston et al. 1996; Firmin-Sellers 1996; De Soto 2000; Haber et al. 2003; Ostrom et al. 2003; Evans 2007; McHarg et al. 2010). Broadly defined, property rights are a set of interpersonal relationships that underpin the ability and power of economic agents to use, manage, transfer or alienate resources, or to take an income or rent from their use. A well-recognized definition says that they are 'social institutions that regulate the use of scarce resources by assigning and enforcing rights and duties' (Eggertsson 2005: 27). Social institutions in turn are defined as the humanly devised constraints and associated enforcement mechanisms that generate incentives, behavior and outcomes in social groups (North 1990). Hence, sector legal frameworks are conceived as social institutions that make rights real.

Sometimes, property rights are also described as a 'bundle of rights'. This is because rights are granted with varying degrees of exclusivity, flexibility, duration, transferability, divisibility and quality of title (Scott 2008). For example, licences may be granted for shorter or longer periods of time, and rights to resources may be transferred at an earlier or later stage of the production process, as is the case with the legal differences between concessions and contractual regimes.

The liberal school

The liberal school of property rights is the most well-known strand of the property rights literature. By emphasizing that the provision of well-defined and secure monopoly rights is a critical 'input' for private investment in the resources sectors, this school of thought has provided the intellectual basis for the sector reform approach promoted during the era of liberal capitalism.

Roland Coase made a significant contribution to this school in the early 1960s. He proposed that 'in the absence of transaction costs self-interest will always guide the members of a society to contract for the establishment of political structures and systems of property rights that maximize national wealth' (Coase 1960). This theorem has been drawn upon to support the view that self-interested economic agents prefer exclusive private property rights and are less keen on rights that they have to share with others. Economic agents are also thought to take better care of their own property, so more exclusive rights would generally be of advantage for society as a whole.

For the development of natural resource governance frameworks, this particular reading of the Coase theorem has had two important consequences. First, private economic agents have been expected to exert bottom-up pressure on governments to shape sector legislation and associated contracts and agreements towards the provision of more exclusive rights. In turn, this strengthens the incentives of private entrepreneurs to channel investments towards a more efficient exploitation of natural resource rents, from which governments will also benefit. Secondly, private economic agents are thought to generally be better suited to develop natural resources and exploit resource rents. This puts the right to exclude others from exploiting natural resource rents at the heart of sector legislation and associated contracts and agreements.

At least three bodies of work have supported these interpretations. The first includes an influential series of empirical studies conducted in the 1960s and 1970s on the development of native private landownership in parts of Canada as well as early mining legislation in the American West (Demsetz 1967; Umbeck 1977, 1981; Libecap 1978). These studies provided the empirical basis for the hypothesis that economic actors who stand to benefit from more exclusive and more precisely defined property rights will mobilize to initiate regime change towards this end. Potential winners will engage in collective action and exercise political pressure to demand such rights. The initial triggers for such demand have been thought to be exogenous and are similar to the external factors highlighted in the introduction of this volume, including global geoeconomic and geopolitical power shifts, technological innovations and demographic and migratory pressures. As these factors lead to increase in the value of resources in the ground, it becomes worthwhile for potential winners to demand regime change and to exploit the new economic opportunities.

The empirical story goes that miners in the American Gold Rush were driven by such opportunities but were seriously hampered by contestation over claims of ownership. So they started to engage in collective action to develop localized practices that would help them to resolve their disputes.

In the absence of a credible public authority stepping in to define and enforce rights, miners collaborated to develop localized mining regimes that assigned and enforced individuals' rights over the subsoil resources that had been explored and discovered. With the introduction of order into the initial scramble for mineral resources, local miners gained greater certainty. This in turn allowed them to become more efficient and benefit from increases in scale. Later on these local district mining regimes were incorporated into US federal mining legislation (Libecap 1996). Some have considered this development the beginning of America's successful path of resource-based economic development (Wright and Czelusta 2007).

A second body of work has investigated the impact of the 'rule of law' on economic performance. Based on the policy conclusions derived from the application of the Coase theorem, a series of cross-country quantitative analyses have demonstrated that countries with stronger 'rule of law', that is, better protection of property rights, outperform those with poorer governance indicators (Haggard et al. 2008). From this research, policy advisors have concluded that countries seeking to boost economic growth by drawing on their natural resources wealth should first and foremost strengthen their sector legal frameworks.

The third body of work evolved around the postulate of the 'tragedy of the commons'. In 1968 Garrett Hardin suggested that when resources are owned in common rather than by private individuals, each user has an incentive to overexploit these resources for his/her personal gain (Hardin 1968). The result is unsustainable resource exploitation, which ultimately undermines the common interests of the community. Thus, ownership over natural resources should preferably be assigned (or licenced) to private economic agents because they would balance their short-term and long-term interests and ensure that resources are exploited in a sustainable manner. This policy conclusion has also been used to argue that granting exclusive rights to private resource developers would be best for ensuring the sustainable exploitation of subsoil natural resources (Daintith 2010b: 5).

This reference to Hardin's postulate is surprising for two reasons. First, Hardin has been challenged for confusing resources 'owned in common' with so-called open access regimes. However, the distinction between these two types of property rights regimes is critical with respect to conflicts between resource developers and rural and/or indigenous communities whose livelihoods and/or group identity depend on resources, which they own in common on the basis of traditional forms of governance. This includes such things as grazing rights, rights to water, rights to cultural sites and other land user rights. For example, the Australian Native Title Law obliges resource developers to follow a process, which establishes whether and what kinds of indigenous rights to resources exist in the licence areas obtained. Second, Hardin targeted his hypothesis on renewable resources. But elaborate empirical and theoretical research conducted since the 1960s has clearly shown that there is not a single form of ownership that constitutes a *per se* first-best solution for sustainable resource governance (Ostrom et al. 2003). Irrespective of whether resources are owned privately, by the state, or by communities, other additional variables are necessary for ensuring sustainable outcomes (Ostrom 2005).

The critique of the liberal school

The liberal school of property rights has dominated in the past, and arguably still dominates, policy advisory circles. But its weaknesses have not gone unnoticed and have led to several strands of research. The critics of the liberal school include a mix of economic historians, institutional economists and lawyers who have investigated how non-market institutions have supported economic development through encouraging local, national and international trade and finance (North 1990; Olson 2000; Williamson 2000; Acemoglu 2003; Acemoglu and Johnson 2005; Eggertsson 2005; Evans 2007; Nicita and Pagano 2008; Acemoglu and Robinson 2012).

These scholars have raised several interconnected issues. First, the liberal school makes the assumption, based on the Coase theorem, that economic decisions and transactions are free of costs. But in the real world there are always transaction costs, for example those associated with obtaining information, engaging in bargaining, and enforcing agreements.[5] Because they have studied the broader historical and social foundations of capitalist market economies, these critics highlight the important role that social institutions play in helping economic agents to reduce and manage transaction costs. They argue that social institutions coordinate behavior and this affects the risks that economic agents associate with the costs and the returns of their economic decisions. They also constrain economic agents in how they can manipulate and alter property rights to their advantage and to the potential detriment of broader social outcomes. Thus social institutions are critical to how capitalist markets work, or do not work.

A second issue is how socially beneficial market institutions have evolved and why they have not evolved everywhere. The argument is that transaction costs advantage some economic agents over others to shape social institutions.

The third issue is that exclusive property rights cannot simply be secured by reforming sector legislation and signing contracts and agreements. The strength of such rights is underpinned by their legitimacy, which is the 'outcome' of political processes taking place largely within the borders of sovereign nation states. The critical function of legitimacy is to reduce the cost of enforcing property rights. The argument is that the legitimacy to grant exclusive rights requires a positive trade-off between the benefits and costs associated with the granting of such rights. A key factor affecting this cost-benefit trade-off is the extent to which citizens are able to challenge public authorities to deliver on the duties and responsibilities of sovereign nation states vis-à-vis their citizens. For example, if governments put in place broadly accepted measures that ensure that potential negative externalities are not simply off-loaded onto local communities and other economic sectors, this reduces the likelihood of protests and other costly forms of discontent.

It should be stressed that these critics of the liberal school do not deny the importance of secure property rights for positive investment decisions, including encouraging private companies to explore and produce oil, gas and mineral resources in countries with difficult political and economic environments. But they stress that reforming a country's sector legal framework to pursue the narrow objective of attracting foreign investment is not enough to ward off and discredit potential domestic resentment against such reforms, nor is it enough to bring about economic development. Ultimately, by focusing only on a narrow set of benefits associated with granting exclusive property rights, the real costs associated with the provision of such rights have been ignored (Pistor 2002; Nicita and Pagano 2008; Katz and Owen 2009). Important political and social implications inevitably arise from how these costs are distributed and this can affect the relationship between governments and resource developers. For example, industry observers have noticed that more governments are now designating responsibilities directly to resource developers, for example in the form of mandated social investments, and/or setting increasingly stringent conditions around specific development objectives such as local content and local employment targets, financing shared infrastructure or delivering subsidized resources to domestic markets.

To delve further into these critical arguments, we can again focus on three questions. The first question is what role do nation states play in the provision of property rights? The answer is that private companies can only enjoy exclusive rights over natural resources, if some sort of collective entity absorbs and distributes the costs of defining and enforcing such rights for them. Nation states play a critical role in defining, guaranteeing and enforcing private property rights, irrespective of who initially owns subsoil natural resources.

Philosophically this answer is underpinned by the argument that nation states hold a monopoly over the use of sovereign power. This gives them a comparative advantage over the provision of property rights. In exchange for more exclusive property rights, nation states can impose duties and responsibilities on the beneficiaries to cover the costs associated with the provision of such rights, including administrative, political, social and environmental costs. When private resource developers are granted rights to explore and develop subsoil resources, they have to share part of the economic rents with the nation state. This is usually called taxation, but may also involve other forms of revenue mobilization. For example, in countries where tax administration is particularly weak revenue may be mobilized in-kind, such as Chinese companies building shared infrastructure in the Democratic Republic of Congo. In theory, countries and resource developers should both gain, as long as the wider socio-economic benefits derived from granting exclusive property rights outweigh the political, economic and social costs of defining and enforcing such rights.

A historical argument also supports this role of the nation state. The argument goes that in expanding their territorial reach across Europe, the ruling elites of emerging nation states needed better access to capital, which prompted them to strengthen their power. Initially, sovereign power was exercised through the deployment of armies paid for by coercively taxing communities. But political elites learnt to use their authority more effectively and collaboratively, by forging productive alliances with new economic elites. The development of mineral law in Europe is part of this story whereby the 'first come, first serve' principle reflected a mutually beneficial deal between political rulers and free miners. Nation states developed increasingly elaborate political-administrative systems to define and enforce exclusive resource property rights for eventual, though not immediate, expansion to broader constituencies. New economic elites created new value, thus giving the state greater access to capital and allowing it to expand its revenue base and eventually provide more public goods and services that further enhanced broader-based economic opportunities (Tilly 1992; Lachmann 2000; Bates 2001; North et al. 2009; Lichbach 2010; Tanzi 2012).

The development of district mining legislation in the American West provides an example of what happens in the absence of a collective political entity as represented by nation states. The previous subsection explained the view of liberal scholars that, in the absence of such a collective authority, local miners engaged in a process of bargaining and negotiating district mining legislation from the bottom-up. By adhering to the Coase theorem, they assumed that this process was free of transaction costs, or at least that transaction costs made no material difference. However, critics argue that in the absence of an authoritative public entity local miners faced a serious collective action problem to solve disputes over competing claims among themselves. What the miners did to minimize transaction costs is to resort to their shared knowledge of the European principle of 'first come, first serve' as a working practice for managing claims.

To test this hypothesis, Clay and Wright (2003) investigated historical records on claim disputes and controversies and found that district mining legislation did not fully resolve conflicts around mining claims. Many incidences of dispossession, favouritism and cronyism in dispute resolution continued to occur even after the bottom-up negotiated district mining legislation had been developed. They conclude that the collective action of local miners did not lead to the development of a minerals regime that provided secure titles. Adhering to the principle of 'first come, first serve' merely helped local miners to establish a system that helped them manage 'access' to claims. This constituted a marginal improvement to the preceding chaotic scramble. Property rights only become more secure with the further consolidation of the liberal American nation state developing and enforcing sector legislation under the 1872 Federal Mining Act.

The second question is what are the wider socio-economic benefits of providing property rights as a 'public good'? Here the argument is that nation states support the creation of value and expand their revenue base, if they are able to define and enforce property rights more broadly than just for a narrow set of beneficiaries (De Soto 2000). By implication, the oil, gas and mining sectors are more likely to deliver wider socio-economic benefits, if such rights are also provided to other economic sectors as part of a comprehensive and consistent legal framework. In more theoretical terms, this is to say that nation states ought to provide property rights as a 'public good'. This is the case when the protection and security of the rights of one economic agent does not negatively impact the protection and security granted to another economic agent and his/her rights. It is this aspect that matters most for countries providing strong 'rule of law' (Acemoglu and Johnson 2005; Hoff and Stiglitz 2005; Katz and Owen 2009). This suggests that when policy advice focused just on reforming legal frameworks specific to the oil, gas and mining sector, this advice was effectively based on a misinterpretation of the findings of comparative research of the impact of legal frameworks on economic development.

The fundamental challenge with public goods is that they tend to be undersupplied. This is because by definition they are 'non-rival' and 'non-excludable': everybody benefits from their existence but nobody wants voluntarily to pay for their provision. As game theory has shown, nation states can address this collective action problem by deploying their monopoly over the use of power. Using taxation they can compel beneficiaries to commensurately pay for the benefits they derive from the provision of exclusive property rights. Once the collective action problem is overcome, economic opportunities expand and more people can benefit. This is what characterizes socially responsible market capitalism. This is of course the empirical story for Europe, where domestic political pressure increased the costs for political elites providing monopoly rights to themselves and their aligned economic elites. The responses of political elites to the costs of rising pressure from new economic elites, as well as from organized labour and other broader-based social constituencies, was critical for maintaining political order and an overall positive trade-off. But this required a commensurately higher level of public benefits. In Europe, this development led to an expansion of administrative capacity to tax and a more broad-based provision of property rights. Thereafter came the transition away from coercive use of power to less person-focused and more rules-based systems of governance. Put together, it ultimately brought about a higher level of state financial and regulatory activities but also greater personal freedoms, while at the same time it scaled back the privileges political elites had previously enjoyed.

In countries with considerable oil, gas and mining reserves it is often the case that secure property rights are not provided as a public good. Political elites define and enforce exclusive rights only to a small constituency, so-called *clubs* of beneficiaries, which generate the revenues that allow political elites to survive. Because the oil, gas and minerals industry can survive without extensive linkages to other domestic economic sectors as long as production inputs can be imported from abroad, political elites can more easily get away with providing property rights as 'club goods' to a select group of resource developers.

A number of comparative case studies on resource-rich countries have used the 'club goods' hypothesis to show how narrow alliances between political and economic elites can limit the provision of secure property rights only to a select group of resource developers, who in turn generate sufficient financial and other benefits for political stability to be maintained (Haber et al. 2003; Snyder 2006; Luong and Weinthal 2010). The consequences are twofold. While narrow political-economic alliances do well, there are limited wider benefits for the economy and society as a whole. Secondly, while 'club goods'-based political arrangements can remain stable for some time, they are destined to run into difficulties when the benefits become insufficient to counter domestic pressure building up to distribute benefits more widely and to compensate those who bear the negative consequences of the country's overt reliance on an export-oriented resources sector.

When faced with growing domestic pressure, governments need to gain domestic strength. This can lead to at least two scenarios. First, governments can increase their use of coercive power to maintain the political status quo. The cost of this strategy depends on the relative strength of non-elite domestic constituencies. Secondly, governments can aim to increase their legitimacy and political strength by expanding the positive reach of the state apparatus to improve public policies and provide more inclusive economic opportunities. It is difficult to dismiss the pursuit of the second strategy as just another feature of 'resource nationalism'. On both normative as well as analytical grounds, nation states should not be undermined when they strive towards expanding the provision of property rights as a 'public good'. Indeed, it is a distinctive feature of liberalism, as identified in Chapters 4 and 5, that it seeks to promote international norms around the duties of nation states to discourage elites from pursuing the coercive route. The challenge for resource developers is that this invariably means higher production costs, and therefore a loss of value as estimated on the basis of the status quo.

The third question is how do elite constellations and pre-existing institutions affect cost-benefit trade-offs? While nation states play a critical role in defining, guaranteeing and enforcing private property rights, this role does not automatically benefit everybody. Nation states are not monolithic entities. It would be rather unrealistic to assume that nation states as collective entities can perfectly recognize all costs, ensure that they collect a com mensurate share of the resource rent that resource developers generate, and re-allocate the rent received to cover administrative costs and compensate the individuals, communities and other economic sectors that have been negatively impacted. In reality, the true costs are often difficult to recognize at an early stage. And when they arise later some can more easily be socialized or off-loaded onto social groups that are too weak and/or too unorganized to avert this outcome. The presence of transaction costs makes it very likely that some agents are more privileged in capturing benefits, while others are disproportionately shouldering costs and may struggle to negotiate some form of compensation or benefit sharing.

Pre-existing institutions facilitate the strategic actions of different constituencies to manipulate trade-offs and to direct institutional changes towards serving their particular interests. Thus negotiated changes to property rights do not necessarily need to present an improvement, at least not for everybody. Yet, with its narrow focus on benefits the liberal school assumes that changes to property rights regimes towards more exclusivity always constitute an overall improvement. Crucially, this ignores the moral question about how much inequity in the distribution of benefits and costs a government can justify. Secondly, it ignores the economic question at what point a skewed distribution of benefits and costs institutionalizes unproductive rent-seeking and other features of the 'resource curse' that undermines a country's overall social and institutional capital.

This critique does not suggest that reforms driven by vested interests are inherently bad. But it does demand that an analytical lens be placed upon the issue and prompts more careful consideration of the domestic distribution of the costs and benefits, rather than just assuming overall positive outcomes. Shaping exclusive resource property rights regimes towards the interests of advantaged elites can serve a country well over the medium to long term, depending on the configuration of additional variables. It is these additional variables that the critics have sought to identify, by contrasting and comparing country cases.

One of the analytical concepts used to identify under what conditions a positive overall trade-off might occur is the 'commitment problem' (Olson 2000; Acemoglu 2003; Bates 2008). This problem is defined as the vicious circle of 'any government strong enough to arbitrate property rights is also strong enough to abrogate these rights'. This means that the power of a nation state capable of defining, enforcing and arbitrating property rights is not by itself sufficient to prevent the malign abrogation of these rights when this suits particular interests. Theoretical solutions to the commitment problem can reveal under what conditions abrogation might, or might not, take place.

An important line of enquiry has been whether, in comparison to unsuccessful cases, there are common configurations of critical variables across successful cases (Eifert et al. 2003; Haber et al. 2003; Snyder 2006). Comparative case studies have shown that positive results are most likely when there are self-reinforcing bargains struck between political and economic elites as well as between these elite alliances and non-elites. This points to two critical relationships: the first is the relationship between political elites and the economic constituencies upon whom government depends for most of its revenues. The second is that between these elite alliances and wider socio-political interest groups upon whom the government relies for its broader legitimacy (but who might also influence the non-government economic elites). It would appear that the more critical relationship is that between the nation state, as the collective entity granting and guaranteeing exclusive property rights over resources, and the wider constituencies putting pressure on how government grants such rights, imposes duties, absorbs costs, collects revenue and reallocates expenditure to serve broader economic, political and social objectives. For an overall positive outcome, the critical factor appears to be achieving a mutually advantageous tripartite dynamic between political and economic elites and their relationships with broader socio-economic interest groups. Exogenous shocks to domestic market conditions can affect this relationship in positive as well as negative ways, not least conditioned by whether resource developers decide to respond to public needs in a confrontational or a more collaborative manner.

To summarize, this section has drawn on the social science concept of property rights to explain the ideas that have shaped sector legal frameworks during the era of liberal capitalism. This included an analysis of the liberal school of property rights as the basis of the reform thinking that emerged from the 1980s onwards, which was then juxtaposed with the views of critics who have looked at the role of property rights from a wider social science perspective. A key insight drawn from these critiques is the way that the assumptions underpinning the liberal perspective have failed to assess more fully the costs, duties and trade-offs associated with granting exclusive property rights specifically to the oil, gas and minerals sectors. Liberal reforms have focused on achieving one specific policy objective, to attract foreign investment, and have uncritically associated this with positive overall benefits. The unintended consequence has been that governments, resource developers and other affected parties have not been alerted to the risks and contingencies associated with this approach (Berkowitz et al. 2003). The lack of caution that sector legal advisors have displayed in translating the work of academic researchers into policy conclusions has put the legitimate needs of government to broaden the objectives of public policy at odds with how resource developers have taken their investment decisions. Once these decisions have been taken resource developers have a strong interest in maintaining the legal status quo, so as not to lose perceived business value. However, the ultimate loss of value in the future could be greater if resource developers and governments subsequently confront each other, instead of exploring new ways of collaborating to ensure broader benefits are delivered to host countries. The final section sets out where the main pressure points are likely to lie in the regime transition from liberal to state capitalism.

Conclusion

The core premise of this chapter is that resource property rights lie at the heart of many natural resource-related conflicts experienced in countries and regions across the world. During the era of liberal capitalism, sector legal frameworks were viewed as a critical 'input' for attracting foreign investment. This idea was drawn from a particular reading of the liberal school of property rights, which suggested that potential winners will push for the development of sector legal frameworks that enable them to exploit economic opportunities and that this would serve society as a whole. A reading of the broader social science literature suggested another way of thinking. This is that property rights are better conceived as the 'outcome' of political processes that take place largely within the borders of nation states, irrespective of whether such processes were initially triggered by external factors. Rather than focusing narrowly on the substance of legislation, contracts and agreements, analysts should rather pay closer attention to path dependencies, transaction costs and the overall legitimacy of property rights regimes. Critics have highlighted a number of areas where the liberal school has ignored important institutional aspects, including the positive role of nation states in defining and enforcing property rights regimes, the provision of property rights as a 'public good' and the impact of pre-existing institutions and dynamic elite constellations on the cost-benefit trade-offs for society as a whole.

This critique has corresponded with a shift in thinking towards recognizing the obligations and responsibilities of nation-states vis-à-vis their citizens, and away from the earlier and more limited focus on the rights of resource developers. While in popular policy debates this development is still interpreted as a sign for the rise of 'resource nationalism' and it is mainly perceived as a threat, this chapter has cautioned against interpreting every government's attempt to introduce a wider set of policy objectives as a matter of conflict around the rebalancing of power between exporting and importing countries. If resource developers were to become more open to understanding the domestic challenges of host countries and stop viewing sector legal frameworks merely as an input to investment decisions, they would begin to identify where there might be room for new forms of collaboration to retain value and protect themselves against political and socio-economic risks.

Large-scale oil, gas and minerals projects (often foreign-funded and focused on exporting the resources produced) offer both opportunities and costs. Socially responsible market economies, which can be considered as the mildest form of state capitalism, allows the opportunities and costs to be assessed and negotiated within the domestic political arena. The alternatives are more authoritarian forms of state capitalism where narrow alliances between political and economic elites are more likely to seek to monopolize opportunities so as to capture rents at the expense of others. In coming to terms with the end of the era of liberal capitalism, international resource developers and the home governments backing them will have to decide what balance between collaboration and conflict they want to deploy to shape the new era towards more socially responsible or more authoritarian forms of state capitalism. Through the power that resource developers can yield upon their home governments, their joint responses abroad will inevitably bear consequences for the way economies will evolve both at home and abroad. Benner et al. (2010) have cautioned that 'at the very least, the powerful can no longer claim to not know the tragic outcomes of bad resource governance'. This chapter has argued that the same applies to viewing property rights merely as 'inputs' to taking investment decisions.

Acknowledgements

The author wishes to thank Ana Elizabeth Bastida, Olle Ostensson and Roland Dannreuther for their helpful comments on two earlier drafts. Possible errors remain the responsibility of the author.

Notes

1. The sector-specific literature usually refers to the resource developer consortium as the 'contractor', 'licence holder' or the 'concessionaire'. Often there is also no distinction made between the resource developer taking resources out of the ground and investors providing funding to undertake this economic activity. Both are interchangeably referred to as 'the investor'.
2. Off shore subsoil natural resources belong to the US government, who also owns a considerable amount of on-shore land, so-called federal land.
3. See in particular Chapter 3.
4. An exception is Columbia where a concession is a contract (categorized as an 'adhesion contract'). Recent modifications to the Colombian mining code entitle the state to reserve areas for large-scale mining and to grant them to private parties under a 'concession contract', similar to the type of agreement used in the oil and gas sector.
5. The Coase theorem has served different strands of economic and legal thinking, including conflicting interpretations of its meaning. See Nicita and Pagano (2008) for a more detailed debate.

Part IV

Scarcity, Technology and Future Supply

9
Peak Oil: Myth or Impending Doom?

Patrick Criqui

The debate over the prospect of a peak in oil production occurring in the near future radically divides oil experts. Some are convinced that the comparison of the total estimated recoverable resources with the quantities of oil already discovered and produced points to the inevitability of a slowdown in production growth, to be followed by a stabilization in the near future and a decrease over the following decades. Conversely other experts, often economists, consider the 'peak oil' prophecy to be a fallacy consistently undermined by empirical evidence, in particular by the slow but continuous increase in global proved oil reserves. Indeed, according to BP, these reserves amounted to 1380 Gbl in 2010 as against 670 in 1980.

According to Colin Campbell, one of the most prominent advocates of peak theory, 'the term peak oil refers to the maximum rate of the production of oil in any area under consideration, recognising that it is finite natural resources, subject to depletion'.[1] The main objective of the Association for the Study of Peak Oil and Gas (ASPO), which he co-founded, is to expose to as many people as possible the 'objective phenomenon' that 'the world's petroleum resources are becoming depleted, to marshal the empirical data that support the concept and probable imminence of a global oil peak and to attempt to understand and quantify the impact of an oil peak on society, the economy, and the lives of ordinary people' (Zhao et al. 2009). The main message of the ASPO community is that the peak is here already – or at least coming very soon – and that the end of cheap oil is imminent.

Even if the exhaustible characteristic of oil and gas is an 'objective phenomenon', critics have put forward a series of strong arguments against the peak oil concept. They argue that the prophets of doom have been wrong for over 60 years and have overlooked the impact of increasing knowledge and improved technologies on the capability of the oil industry continuously to recreate the reserves that are destroyed in the hydrocarbon production process. Combined with this increase in knowledge – and also encouraging it – the rise in oil prices since the period of the oil shocks has strongly stimulated the continual economic renewal of this ultimately limited resource.

Two subjects are thus at the core of the controversies: the first one is, of course, when will the world face a levelling-off in oil production; the second one is what will be observed after production achieves its peak, will there be a rapid decrease or will a plateau be reached?

In this chapter we propose an 'agnostic' approach to the peak debate, while highlighting the complexities of determining the facts and the quantitative evidence. The often diverging empirical forecasts can only be reconciled when focusing on the dynamic aspects of the oil discovery and production process. This obliges one to consider the two sides of the 'tug-of-war between diminishing returns and increasing knowledge', which, according to Adelman (1993), is the very essence of the economics of natural resources. The outcome of the tug-of-war is never predetermined and our diagnosis is that although the peak oil theory may be wrong in its assessment of the timing of the peak and on the shape of the decline thereafter, one may expect a growing gap, as far as conventional oil is concerned, between the dynamics of production expansion and the growth in the demand for cheap liquid energy carriers, most notably for automotive transport. While it turns out to be more and more difficult to find and produce conventional oil outside the resource-rich OPEC regions, the huge potential demand from the rapidly growing emerging regions of the world will impose strong pressures on the international oil market in a near future and encourage the development of non-conventional oil with, in most cases, higher economic and environmental – both local and global – costs.

The purpose of this chapter is thus to review the concept of peak oil, critique its main propositions and assess the arguments advanced by oil optimists against those of peak oil. The chapter is structured as follows. In the second section, we begin with a presentation of the Hubbert peak theory and of some recent applications of the theory at the global level. We then introduce a revised conceptual analysis and set of variables with a dynamic analytical framework. The third section assesses the issues behind the polarized debate between the peak community and the oil optimists. In particular, the methodological shortcomings that oil optimists consider as fundamental flaws of the peak theory are highlighted. In the fourth section, we examine the recent attempts by modellers and forecasters to overcome these shortcomings, which include more precise information about the different resources and particularly the taking into account of non-conventional resources. Finally, we examine these issues taking into consideration the potential impacts of international climate policies. In particular, we identify scenarios where sufficient implementation of mitigation policies at the global level would potentially change the very nature of the peak oil problem, turning it from a peak supply to a peak demand problem.

The Hubbert peak and the dynamics of the 'creaming curve'

The Hubbert peak

Marion King Hubbert, a US geologist, was the founder of the 'peak oil theory' as he announced at an American Political Institute (API) meeting in 1956 that, given the bell-shaped curve of the production profile of individual oil regions, US production (lower 48)[2] would reach its maximum by the early 1970s, when almost half of total resources would have been produced. The reasoning behind this was simple: the basic assumption is that the production profile of any oil-producing region follows a bell-shaped curve; the maximum of production necessarily corresponds to the point when about half of the total recoverable resource has been produced. The original prophecy was indeed verified as US production peaked in 1972 and only the development of Alaskan fields allowed some rebound up until 1985, when a new decline followed. A recent study by the Australian government on the future of oil production confirms the validity of the Hubbert curve, at least for the US lower 48 region, from the start of production until 2010 (Figure 9.1).

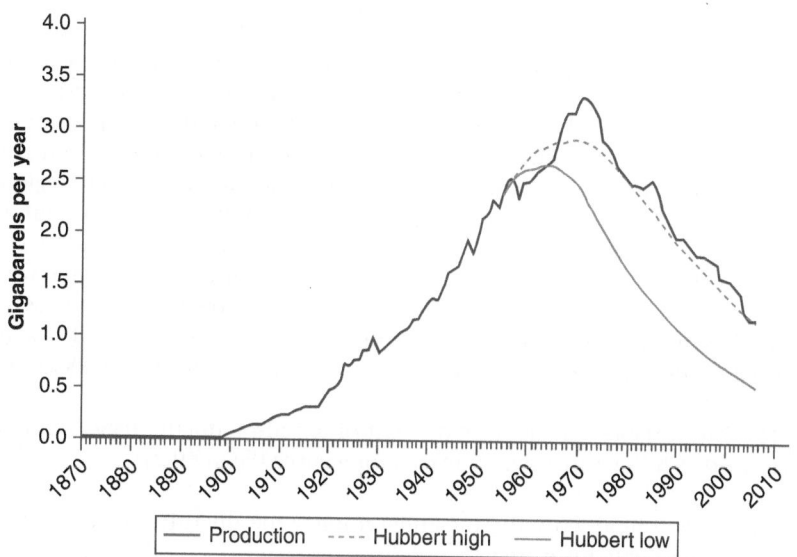

Figure 9.1   Hubbert's (1956) curves and actual US lower 48 production
Note: Data does not capture the surge in US production after 2008 due to shale oil production.
Source: Bureau of Infrastructure, Transport and Regional Economics, Transport Energy Futures (BITRE) (2009) Long-Term Oil Supply Trends and Projections, Report 117, Department of Infrastructure, Transport, Regional Development and Local Government, Canberra, Australia.

After preliminary extrapolation of Hubbert's theory to the global scale in the 1990s, ASPO was founded in 2000 by Colin Campbell in order to develop the approach for the analysis of oil-production dynamics in different regions of the world and by aggregation, to provide a global assessment. The well-known principal conclusion and most striking statement of ASPO[3] is that peak oil is upon us and that the end of cheap oil is there, or almost there (Campbell and Laherrère 1998). Depending on different authors, members of the ASPO forecast that the peak is already here or that it will occur before 2020. One of the key arguments set forward is that annual net real discoveries, to be clearly distinguished from existing fields' reappraisals, have become inferior to total yearly production since the beginning of the 1980s, which confirms the argument of the peak and the inevitable decrease of total oil production as the reservoir from which production is withdrawn necessarily shrinks.

The peak oil theory is a phenomenological approach to long-term oil production profiles, with a simple mathematical basis and a limited number of variables, that is, mostly resource stocks and production flows. However, this simple theory should be replaced by a more complex description of a dynamic context, with causal relationships between different variables in the so-called oil discovery process models.

The dynamics of the 'creaming curve'

Many of the discovery process models are based on the use of 'creaming curves'. This is the key concept used by the oil industry to characterize the dynamic process of the development of an oil-producing region through exploration. The creaming curve describes the decreasing returns between the exploration activity and the total discoveries for one given region. When expanded by including the limits imposed by the ultimate resources available and by the impact of technological progress in oil recovery technologies, it can provide a dynamic analytical framework. This framework organizes in a clear causal system the different concepts that are necessary to understand the dynamic development of oil or gas. Such a causal system is used for instance in the oil discovery process module of the POLES model (Kitous et al. 2010).

The key variables in the development of an oil-producing region can be described as follows in a dynamic creaming curve (Figure 9.2):

- *Oil in place* represents the oil underground, the quantity of which will never be exactly measured.
- *Ultimate Recoverable Resources (URR)* are derived from oil in place as the quantity that can be extracted under current technological conditions. When changes in recovery technologies are taken into account it is more adequate to use the term 'recoverable resources', a quantity that is indeed increasing with technological progress (i.e. enhanced oil recovery (EOR)).

- *Cumulative discoveries* correspond to the oil that has been discovered since the very beginning of the oil exploration process in the region considered; by logical consequence, cumulative discoveries are inferior to the recoverable resources.
- *Reserves* are finally calculated as the difference between cumulative discoveries and cumulative production since the origin.

It is clear from Figure 9.2 that, given the different boundary conditions, the dynamics of reserve formation may allow for a full reproduction or even increase of reserves during a long period of activity, even for a mature region. However, it also shows that given the decreasing returns of the discovery process, reserve addition will eventually decline and become inferior to the annual production level. After that point, reserves and the reserve/production ratio will decline, resulting in the decrease in production. In that sense the dynamic creaming curve described above is fully compatible with the Hubbert curve for a producing region in that it describes similar production profiles in a series of causal relations. Furthermore, it provides a satisfactory answer to one of the most common critiques of the peak oil theory, as it takes into account technological progress in oil recovery technologies.

Cumulative production, reserves and discoveries: Assessments and reassessments

At the time of the creation of the ASPO in 2000, the association estimated the URR at 2000 Gbl or a little bit less, while cumulative production was 870 Gbl and yearly production 27 Gbl/yr (see Table 9.1). At these production

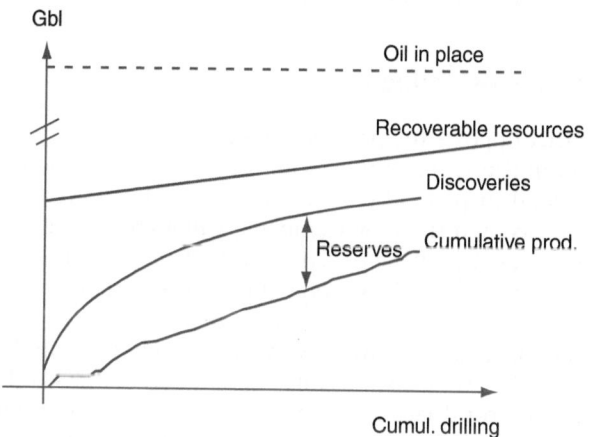

Figure 9.2 Oil in place, recoverable resources, cumulative production and reserves
Source: POLES model.

Table 9.1 Cumulative production, reserves, discoveries and resources revisited

Gbl	2000	2010
Yearly production	27	30
(1) Cumulative production	870	1,160
(2) Remaining reserves (BP)	1,100	1,380
(1) + (2) Discoveries	1,970	2,540
Ultimate recoverable resources	1,800–2,000[a]	3,000–3,500[b]
(1)/((1) + (2))	*44%*	*46%*

[a] ASPO estimate.
[b] EIA estimate, conventional oil.

levels, the mid-point would be reached within five years and this led to the prediction of an imminent peak towards 2005. Ten years later, cumulative production was indeed 1160 Gbl with a yearly production of 30 Gbl, but the URR have been considerably revised upwards, at least by the International Energy Agency (IEA), which had a maximum estimate of 3500 Gbl. On the IEA assessment, with a slowly increasing annual production, the mid-point would be reached within 20 years, that is, before 2030.

Over ten years, the dominant view of the experts who consider the stocks and flows aspect of the oil problem has shifted from the vision of an imminent peak to that of a pending problem that will have to be managed in the medium term. But not all experts consider this approach as valid, as most economists emphasize the role of economic and not physical variables in the changing picture of oil development.

Economists versus geologists

The debate that opposes the peak oil community (a majority of geologists) and the oil optimists (a majority of economists) can be synthesized from the arguments found in papers by Adelman, Odell, Mabro, Yergin and Lynch (the oil optimists) and those by Campbell, Laherrère, Bauquis and Aleklett (the peak oil theorists). The application of the Hubbert curve at world level is the extension at a 'macro' level of a 'micro' level phenomenon and looks quite powerful. Nevertheless, Mabro (2006) points out that to move beyond the merely tautological – oil and gas are exhaustible and will therefore one day be exhausted – two questions must be carefully addressed. The first one is when will the world face a levelling-off in oil production and whether we are years or decades away from this point; the second one is what will happen after production reaches its peak – will there be a steady decrease or will a plateau be reached?

The peak: When and with what consequences?

The controversy is all about answering these two questions. Of course a set of major uncertainties are at the heart of this controversy and the arguments between the peak oil community and the oil optimists derive from these uncertainties. Beginning with the ASPO members, one can note different degrees of pessimism among them.[4] For some, the global oil production has probably already passed its maximum. That implies that we have reached the peak of the oil age and of cheap and abundant oil (Aleklett et al. 2010). Others consider that the peak is more likely to occur later, that is, around 2020 (Babusiaux and Bauquis 2007). In contrast, the oil optimist institutions or economists, such as CERA, Odell or Lynch do not really consider that the limitation of oil underground will be the core of the problem during the next two decades. They argue that there will be no physical or 'underground' problem and that no decrease of production is likely to be observed at least in the medium term (Odell 2003; IEA 2008; CERA 2009; Lynch 2009).

Regarding the pattern of production that will be observed after production reaches its maximum, peak oil theorists believe that an absolute peak will occur, followed by a steep and irreversible decline. In contrast, the oil optimists see a succession of temporary peaks. As such, production would be able to recover after such temporary peaks and price hikes, when technology is improved and new fields are discovered. For most economists, the scenario of an 'undulating plateau', characterized by a succession of supply crunches, is the preferred image. One of the corollary implications of the 'undulating plateau' is that the peak will not necessarily imply a sharp and definitive price increase, but rather a regime of instability with price shocks followed by countershocks.

Beyond the differences in estimates of recoverable resources underlying these different visions,[5] one needs to analyse the main arguments of the peak oil community and of the oil optimists. The fact that the stories of doom articulated by the peak oil advocates have all proved to be wrong in the past underpins the scepticism of the economists. For most of them, the peak oil theory is at best a tautology and at worst shapes the debate in a wrong way. Adelman is a clear representative of this stance. He stresses that the exhaustible characteristics of oil and gas is actually a non-problem (Adelman 1990, 2004). The problem is not the question of the ultimate amount of oil in place that sooner or later will run out: oil is a commodity among others. Rather, it is all ultimately about demand, supply and prices. One day, there will be no demand at the price that will cover exploration costs and oil will be left in the ground. As Sheikh Zaki Yamani memorably stated, 'the stone age didn't end for lack of stones'.

From this perspective, the appropriate answer to the question of 'when will the last drop of oil be produced?' is never. According to Adelman, the real oil problem lies with the cartel of OPEC, which involves a distortion of oil economic fundamentals and closes the door for exploration in the most resource-rich regions of the world. Therefore, he considers the fact that new discoveries are inferior to current production level does not represent a convincing argument for the peak oil thesis.[6] The fact that increases in reserves come mainly from new estimates of brown (i.e. existing) fields is not a surprise, as investment in the production stage is simply less expensive than investment in the exploration stage. Indeed, the Middle East countries that offer the best perspectives for new discoveries are also less extensively explored. For Adelman, this is another sign that, due to the OPEC oil cartel, the world oil industry is simply walking with its head down blind to what is going on around it.

For Mabro (2006) and the CERA (2009), the peak oil theory in its simplest form indeed appears as a mere tautology, which is undermined by several shortcomings in its methodology. The main one lies in the static perspective adopted. It does not capture the dynamic process of the development of reserves which is triggered by economic forces and the complex interaction between economic variables (Lynch 1998, 2009; Odell 2010). As one CERA report puts it, 'Hubbert's approach fails to account for fluctuation in demand, technological advances, and the discovery of new hydrocarbon physicals' (CERA 2009). Peak oil theorists are also accused of not taking into account the impact of higher oil prices that are triggered by temporary scarcities. Of course adjustment is imperfect and the response lag is an issue of critical importance but, according to optimists, adjustments are surely going to take place, in terms of reduced demand, technical progress and eventually a rebound in supply.

The role of technology

Technological progress, for which the peak oil community seems to have limited faith, is potentially the key issue. Technology is the 'great multiplier' and the oil industry has always demonstrated a great ability to develop and implement innovative technologies. According to Adelman (1993), the history of any mineral industry reflects 'the endless struggle of nature versus knowledge'. So far 'knowledge' has won the battle. Progress in technology allows for both new discoveries and the increase in the recovery rate through improved technologies that turns non-recoverable or hypothetical resources into recoverable reserves. This is clearly sketched in the McKelvey diagram, which defines reserves as the part of total resources that is both sufficiently well identified and producible under current economic conditions, and which shows that price increases will both stimulate exploration and alleviate the economic constraints. This is why recoverable reserves is an ambiguous concept as it depends both on economic and technological variables. Also, one of the important subjects of debate is the potential contribution of non-conventional resources to future production (see below). Some works from the ASPO and from other institutions have indeed updated their forecasts of future oil production taking into account non-conventional oil and considering both crude oil and other liquid fuel production.

The second major area for innovation in the oil industry, after seismic and exploration technologies, has been the increase in the recovery rates, region by region and globally. It is commonly considered that the average recovery rate is currently around 35 per cent. Of course, this mean value hides large differences between oil fields and regions. But, in any case, what is striking is that even a small increase in the recovery rate adds considerably to accounted reserves even without real new discoveries. Optimists consider that the recovery rate could be as high as 60 per cent 50 years from now. A more conservative view shared by ASPO members is that the recovery rate is unlikely to be higher than 45–47 per cent in 2050. Interestingly enough, recent work by the IEA and Aleklett present a convergent view on the probable contribution of EOR to new production capacities with an extra 6–7 Mbd production by 2030.

A peak demand in place of a peak supply?

Forecasts regarding the dynamics of demand also face huge uncertainties. As noted by Porter, 'experts did poorly on estimating future supply potential and as bad or even worse on the demand side' (Porter 1995). The background to the uncertainties in forecasts and scenarios lies in the time lags for demand adjustments, technological progress for the development of substitutes, and the magnitude of environmental policies. This raises the prospect of a 'peak demand'. This idea is mainly proposed by those who believe that supply will permanently be in excess of demand (Odell 2003; Shell 2008; CERA 2009). From this perspective, 'oil's future seems likely to become demand-side limited so that potential supply-side (resources) limitation represents only a low probability prospect' (Odell 2003).

When considering the demand side of the problem one has to recognize that in most OECD countries oil demand peaked in 2005–6 and, due to the combined effects of the economic crisis and high prices, has subsequently significantly decreased (–8 per cent for total OECD demand in 2010 compared to 2005). This may accord some credibility to the thesis of peak demand. But this ignores the tremendous increase in oil consumption in the non-OECD and non-FSU regions in recent years, which has grown 46 per cent from 2000 to 2010. This surge in oil demand is of course mostly explained by ongoing increases in car ownership. As argued by Dargay et al. (2007), these changes will only accelerate with rapid economic growth to 2030. According to their model, between 2002 and 2030 the car equipment rate – currently at more than 800 vehicles for 1000 habitants in the US – may rise from 16 to 269 in China, 17 to 110 in India and 121 to 377 in Brazil. One can easily imagine the consequences of such trends on gasoline demand in the emerging and developing regions of the world. Clearly, with this level of growth, the peak demand hypothesis does not hold at the global level over the next few decades unless stringent emission reduction policies are set in place.

Despite these projections, the oil optimists maintain their faith in the necessary supply-side adjustments being made by increased investment and improved technologies as a response to price increases. Technological progress, investments in exploration and production and demand-side adjustments are the key issues. These are opposed to the static and determinist perspective of the peak theory. The reality of the complex and dynamic interconnections between economic variables makes the Hubbert method a poor tool for forecasting the timing of the peak. For some, this methodological flaw explains why the members of the peak community have been constantly forced to revise their forecasts for a peak and always to project it further into the future (Lynch 1998).

Non-conventional resources: The new frontier?

There is no strict definition of the concept of non-conventional hydrocarbons and the concept mostly relates to extraction method. According to IFPEN (2012), 'in the case of non-conventional hydrocarbons, the objective is to produce hydrocarbons that are very difficult to extract, either because they are located in beds of very low permeability or because their very nature makes them difficult or impossible to move'.

For some authors the very notion of non-conventional oil should be questioned, as the frontier between conventional and non-conventional is not always clear-cut. This is particularly the case for offshore, deep offshore and ultra-deep offshore resources. However, beyond the differences in production costs, there is one clear reason for maintaining the distinction between conventional and non-conventional oil: it relates to the environmental impacts of the extraction methods for non-conventional oils, which are generally much more damaging both at the local and global level.

Non-conventional resources and anticipated production

Maintaining the essentially geological peak oil approach, some recent works propose updated and detailed analyses that take into account the assessment of non-conventional oil and gas resources, mostly extra-heavy oil, tar sands, oil shale and more recently shale oil, as their potential contribution to future production.[7] For example, Masset (2009) provides an updated appraisal of conventional and non-conventional resources. He first identifies 2650 Gbl of recoverable conventional resources (including 300 through EOR and 350 yet to be discovered), that is, two-thirds of the 4000 Gbl of oil in place. To these conventional resources, 600 Gbl of non-conventional oil should be added, representing only 20 per cent of a total accumulation of about 3000 Gbl. The total 3200 Gbl of recoverable liquids are nearer to the

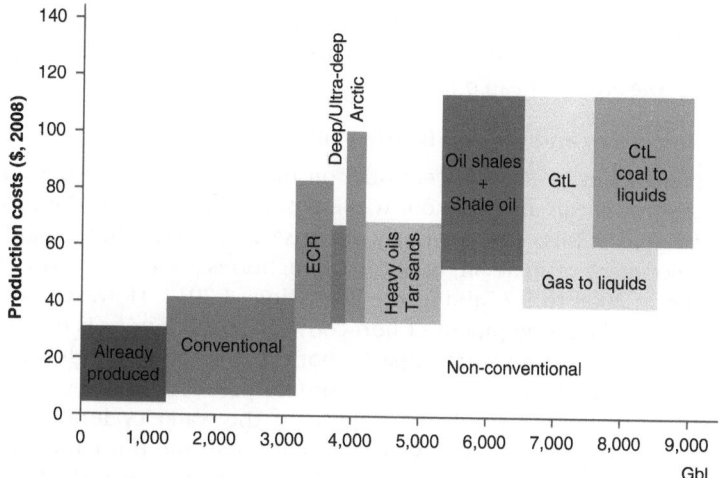

Figure 9.3 Conventional and non-conventional liquid hydrocarbons
Source: IFPEN (2012).

conservative ASPO figures than to the IEA estimates that propose a similar amount for conventional resources alone.

The most recent estimates by the IEA (2008) and IFPEN (2012) of the different resource categories combine the geophysical approach with information on the expected production costs. Their results are much more optimistic. The IFPEN estimate in Figure 9.3 indeed points to global resources superior to 3000 Gbl at a cost inferior to 40 $/bl, 5000 Gbl at a cost of 80$/bl and finally 9000 Gbl at a cost of 120 $/bl (including Gas To Liquids and Coal To Liquids).

As far as anticipated production is concerned, the IEA projects a non-conventional oil production increase from 1.7 Mbd in 2007 to 8.8 Mbd by 2030. On behalf of ASPO, Aleklett et al. (2010) forecast that non-conventional oil could contribute up to 6.5 Mbd by this date. Both IEA and Aleklett et al. (2010) do not foresee a significant contribution from non-conventional oil compared with conventional oil. The POLES model Baseline projection in the FP7 SECURE project provides a projection for 2050, with total oil peaking in 2030 at 96 Mbd and non-conventional oil adding 6 Mbd, and respectively 88 and 10 Mbd by 2050 (Criqui and Mima 2012).

Unsurprisingly, a more optimistic outlook is presented by CERA (2009), Odell (2003) and Aguilera et al. (2009). CERA indeed considers that unconventional liquids already contribute to approximately 14 per cent of total global capacity and predicts that this share will grow to 23 per cent by 2030. For Aguilera et al. (2009) the most important conclusion to draw is that the potential of non-conventional oil undermines the idea that the peak of conventional oil will necessarily lead to a sharp and definitive increase in prices or to the 'the end of cheap oil'.

Replacing cheap and easy oil by dirty oil

The development of non-conventional oil may indeed modify the impact of the end of cheap and easy oil, while ensuring a revival in the production of liquid fossils. This is clearly the case in the US where, thanks to the development of shale oil, total oil production is again on the rise, from 4.95 Mbd in 2008 to 5.7 Mbd in the beginning of 2012. However, one key concern with the development of non-conventional resources is the importance of their environmental impacts, both at the local and global level. These include, first, the physical footprint of production and the impact on local ecosystems; second, the impact on the water cycle in terms of process water requirement and waste water disposal; and third, the increase in GHG emissions due to the production process, treatment and refining of the primary products.

As far as the physical footprint is concerned, the impact on natural ecosystems can be highly visible and obvious as in the case of bitumen produced from tar sands in the Alberta province of Canada. Photographs of the Athabasca basin clearly illustrate the damaging impact on the neighbouring peat lands and boreal forests. Land reclamation might be undertaken by oil companies, but studies show that this reclamation does not prevent a massive loss of peat soils and the corresponding additional carbon losses from these soils (Rooney et al. 2012). For shale oil however, the extensive use of horizontal drilling somewhat limits the aboveground impacts as one well can produce oil from a large underground surface.

The issue of water use and waste water or contaminant products disposal is closely associated to the hydrofracking technology that is common to shale oil and shale gas production. The United States Geological Survey (USGS) is currently conducting a comprehensive research programme to analyse the environmental impacts of shale gas.[8] Recently, the European Parliament commissioned a study on the environmental impacts of shale gas and shale oil, drawing from the US experience but with a specific perspective on the potential development in Europe.[9]

Finally, the GHG content of non-conventional oil is a complex issue as it should be placed in a life-cycle analysis perspective. The results of different studies gathered by CERA[10] point to a Well To Retail pump (WTR) GHG content of the product from oil sands that is 40–70 per cent higher than the average conventional fuel in the US. This however translates into a narrower gap of 5–15 per cent higher emissions on a Well To Wheel (WTW) basis, as most of the emissions occur at the combustion stage.

As can be seen from the above, the question of the environmental impacts of non-conventional hydrocarbons is a complex and multidimensional issue as it incorporates both local and global impacts. One key issue for oil and gas policies over the next decade, and for different regions of the world, will be to decide whether it is worthwhile to engage in the development of non-conventional resources. This will require balancing the advantages in terms of energy independence and reduction in the cost of energy supply and the disadvantages and costs in terms of the local environment and a sustainable climate policy. The initial responses of different world regions are already diverging, with North America clearly on its way to ensuring greater energy independence through non-conventional resources, while Europe, where social acceptability problems appear to be more critical, is till now on a more cautious path. The future of oil over the next decades will to a large extent depend on the policy pattern that will be set in North America and in Europe and, it goes without saying, also in emerging countries like China.

The future: Supply crises or self-imposed sobriety through climate policies?

A common failure among both oil-optimists and -pessimists is their tendency to focus just on 'underground' and/or economic variables. Both approaches leave aside questions of institutional barriers to investment or geopolitical constraints on the activities of the international oil companies. These factors are now the focus of new research dealing with a supply/investments crisis. This has mainly been developed by certain optimists who consider that resource exhaustion is only a fiction, at least for the next few decades (see IEA 2008; CERA 2009; Stevens 2009). The basic insight is that the evolution of supply is going to be mainly shaped by 'aboveground' institutional or political considerations. They argue that supply crunches will probably occur but that these crunches will not have anything to do with the underground availability of resources, but rather with a mix of political, institutional and financial barriers to investment.

Aboveground and underground constraints
In its 2008 report, the IEA argued that,

> capacity addition from current projects decreases after 2010...The gap between what is currently being built and what will be needed to keep pace with demand is nonetheless set to widen sharply after 2010. The sheer scale of investment needed raises question about whether all of the additional capacity we project will be needed will actually occur. If actual capacity addition falls short of this amount, spare production capacity would be squeezed and oil prices would undoubtedly rise – possibly to new record high.
>
> (IEA 2008; see also Forbes 2010)

The key insight here is that institutional barriers to investment and geopolitical issues may be more important for shaping the pattern of future production than the purely geological question of oil availability.

As far as the question of the ability of supply to meet future demand is concerned, this scenario points to some convergence in the forecasts by usually entrenched optimists and pessimists. First, it gives credit to the prospect of an 'undulating plateau' during the next two decades mostly because of the inadequacy of global investment. From this perspective, the world is going to face some impeding supply crises, not due to lack of oil but because of the inability to convert underground resources into productive capacities. Second, these approaches stress that the world oil market is going to remain tight, giving strength to the idea that the inability of supply to meet demand may increase in a shorter-term horizon than the time frame usually adopted by oil-optimists. Whether peak or plateau, the expansion of oil supply will encounter limits over the next decades, thus posing the problem of adjustment in conditions of still powerful potential demand.

In this respect, the analysis provided by P. R. Wells in his twin 2005 articles on oil-supply challenges for the *Oil and Gas Journal* still remains valid (Wells 2005a, 2005b). He argues that

> the world faces challenges rather than impending doom with oil supply. The challenges include a sequence of supply crises likely to develop not when oil production peaks - the subject of much recent controversy - but earlier, when widening gaps appear between demand and sources of supply upon which the world has come to rely.

In his vision the challenges will be organized in a sequence of three stages: Stage 1 with the peak and decline in non-OPEC production; Stage 2 with the reduction in OPEC spare capacity; and Stage 3 with the incapacity of OPEC to meet incremental demand.

A close examination of oil-production profiles in the main oil-producing regions indicates that Stage 1 might be already taking place: from 1965 to 2002, non-OPEC production (excluding US and the former Soviet Union) multiplied by seven, from 4 to 28 Mbd, that is, a 5.4 per cent per year increase; then production peaked and was 26 Mbd in 2010 in spite of the price hikes since 2002. With total OPEC capacity of about 38 Mbd, OPEC spare capacity has also been significantly reduced in recent years (Stage 2) and only the financial crisis starting in 2008 has allowed rebuilding some spare capacity margins. When the crisis ends, the key question will be the ability of the oil regime to facilitate the development of new capacities in the Middle East in order to meet the burgeoning demand of emerging countries. If the pace of this new capacity development is insufficient, then the world will enter into Stage 3 of Wells' analysis.

The problem is thus a combination of underground conditions, that is, difficulties in developing new conventional oil provinces, and aboveground obstacles, that is, the constraints imposed by political and institutional barriers to the development of new capacities in oil-rich regions. Understanding that is important for policy purposes. However, the very nature of the oil problem over the next decades – whether constrained 'underground' or 'aboveground' – will mostly depend on the intensity of the climate policies that will be adopted through international negotiations on the climate change regime.

The impact of climate policies: A downstream solution to upstream problems?

The road towards an international climate agreement – from Rio-1992 and the UN Framework Convention on Climate Change to the roadmap for a global climate pact in 2015 agreed in Durban 2011– has been a long and difficult one. Up until now, the attempts to define a global scheme for establishing national targets consistent with the goal of limiting climate change to an increase in average temperatures of +2°C compared to preindustrial situation have proven unsuccessful.[11] However, the scientific analyses developed and organized in the framework of the four IPCC assessment reports currently available provide an increasingly convincing set of arguments for the reality of climate change, the anthropogenic nature of this phenomenon, and the necessity to curb emissions at world level. The 2°C target would in particular imply a reduction of world emissions by a factor of two in 2050 compared to 2000.

As CO_2 emissions from energy activities through fossil fuel consumption represent more than three-fourths of global greenhouse gas emissions, it is clear that any international agreement nearing this target will have major impacts on the future of energy. Many scenarios exist to describe the impacts of carbon-constrained energy futures, such as those produced by the IEA in its World Energy Outlook and Energy Technology Perspectives, or those developed in the academic community through different sessions of the Energy Modelling Forum (Weyant et al. 2006). Studies performed in the framework of European research framework programmes have also explored these dimensions of the impact of carbon constraints on energy development.

The EU-funded SECURE and POLINARES projects[12] illustrate, based on simulations with the POLES world energy model, the energy consequences of different degrees of emission constraint. In terms of differing scenarios, in the 'No Policy' (or Baseline) case, the future oil development pattern is one of a strong demand from emerging regions and a resulting supply and demand balance that leads after 2030 to the 'undulating plateau'. As mentioned above, this plateau corresponds to a production level of about 96 Mbd for conventional oil, to which 6 Mbd of non-conventional oil are added (Criqui and Mima 2012). In 2050, production is at 88 Mbd for conventional oil and 10 Mbd for non-conventional.

By contrast, the scenario called 'Global Regime', where there is full global compliance with the 2°C target, offers a significantly different picture for the future of fossil fuels. Coal is of course most impacted with a major reduction in total consumption. The impact on oil is less pronounced but nevertheless the world production profile changes significantly: the peak comes sooner in time at 90 Mbd in conventional oil and 4 Mbd in non-conventional oil by 2020, and production is reduced to a level of 57 and 4 Mbd respectively in 2050; that is a 38 per cent reduction from the no-constraint case. This means that the 'peak demand' concept, which as noted above is irrelevant in the absence of environmental constraints, fully applies in the case of the implementation of a strong climate regime. Furthermore, the lower demand profile heavily reduces the development of non-conventional oil resources (by 60 per cent), because they represent the more costly marginal resources.

This does not imply a judgement on the probability of the establishment of such a severe regime, which is another debate. But it is necessary to point to the fact that if there were a regime, and an effective one, then the self-imposed sobriety in the oil-consuming countries would profoundly impact the nature of the uncertainties pending on oil supply. While underground conditions will matter more in the 'No Policy' scenario, the 'Global Regime' would significantly reduce and even fully negate the underground risks for oil supply, leaving room only for aboveground and underinvestment risks. In that perspective, one can of course expect that oil-exporting countries will adjust their investment policies in order to fit to the new demand pattern, but the probability of severe and permanent price hikes would be significantly reduced.

One can also here identify the potential double dividend of climate policies: beyond the often-mentioned and discussed macroeconomic benefits of the recycling of the carbon tax (or emission permit auctions), strong climate policies would significantly reduce the energy vulnerability of the European energy system and, if shared by a sufficient number of countries, also significantly reduce the risks of oil shocks and price hikes themselves, while preserving the earth's climate.

Conclusion

According to economists, the main shortcoming of the peak oil theory is its static methodology. It fails to recognize the dynamic process of reserve evolution that is triggered by economic forces and the capability of the oil industry to develop new resources in a competitive way. Contemporary forecasts try to take into account the dynamic links between economic variables and the impacts of new resources and technologies, thus introducing more refinement in the treatment of the problem.

However, the recognition of these new elements does not entail naive optimism. An 'agnostic' view to the problem of future oil production obliges one to recognize both the limits of the pure peak oil approach and the fact that, due to the nature and strategic importance of this natural resource, it would be unwise to consider that supply will meet demand in any future circumstance. If steep peak oil is no more the most probable hypothesis, there is still a high probability of some impending supply crises due to the expected relative dynamics of oil supply and demand. While the peak oil theory was supposed to identify the principal danger threatening the future of the world energy and economic system, it now appears, when taking into account of the climate change problem, that the promotion of a sustainable world energy system is a much more complex issue than simply ensuring either the adequate level of investment in supply or, contrarily, the 'transition away from oil'.

First it appears that, at least for the short- and medium-term future, the 'aboveground' conditions for ensuring access to resources and adequate investment plans is of critical importance since, in many major oil-producing countries or regions, oil supply is constrained by political will or political instability more than by resource scarcity. Improving these conditions may significantly alleviate market tensions and permit a return to reasonable price levels, which would reduce the pressure on the world economy. The policy prescriptions promoted by the peak oil theory may thus be in contradiction with the need to focus on the financial, political or institutional barriers for access to conventional resources for IOCs and on the political depletion rate chosen by the producer states and consequently by NOCs (see similar conclusions in Chapters 10 and 11).

Second, with current or future price levels near to $100 a barrel, huge resources of non-conventional oil become cost-effective if the cost of carbon is not accounted for: a doubling of total available resources seems a reasonable estimate. The possibility of mobilizing lower-grade resources, as has been the case for most minerals in the past, may completely change the perspective of the future liquid hydrocarbon system. But this would be a move from Charybdis to Scylla as the environmental consequences of non-conventional hydrocarbons development are major, both at the local and global level. In the transition from the cheap and easy to dirty oil, the scarcity problem may thus only be solved at the cost of increased environmental damages.

In particular, and this is the third point, the climate impacts of a massive development of non-conventional resources are potentially huge. This is first because new sources of underground carbon would be made available. These sources require a high level of energy, and thus of CO_2 emissions, in their own transformation process, making them more similar to coal than to conventional crude oil from this perspective. These new resources may also provoke a shift in the energy strategies of many countries that benefit from important non-conventional resources, from the US to France, China or Poland. The shift would compound the temptation to solve the energy security problem not through more efficiency and low-carbon sources such as renewable or nuclear energy, which would be consistent with climate policies, but through an increased reliance on domestic fossil sources.

In the near term, the particular characteristics of the international and domestic institutional framework governing relations between states and operators will mainly drive the future of oil production (see Chapter 1). But from the medium- to long-term perspective, there should evolve a system of careful and long-term management of fossil fuel resource stocks and production flows to be discussed between the consuming and producing countries which also takes into account the global GHG atmospheric concentration constraint.

Acknowledgements

The author wishes to thank Sylvain Rossiaud (EDDEN laboratory) for his contributions to an earlier version of the chapter.

Notes

1. See ASPO's website: http://www.peakoil.net/ [Accessed 26 October 2012].
2. Lower 48 refers to all the territorially contiguous states in the US that is, all states except Alaska and Hawaii.
3. Or the 'Peak oil brigade', as ironically set out by Helm (2011).
4. For a presentation of the different projected dates of the peak from several authors and organizations, see de Almeida and Silva (2009) and the thorough comparative work of most of the contemporary forecasts for global conventional oil production undertaken by UKERC (2009).
5. Recent estimates of ultimate recoverable resources for conventional oil vary within a wide range of 2000–4300 Gb. All methods and assumptions (especially the narrow/wide definition of conventional oil) have been criticized for underestimating/overstating the ultimate recoverable resources (IEA 2008; Aguilera et al. 2009). One of the main conclusions of the UKERC's comparative study is the fact that most of the differences between the two groups of models, those displaying a peak before 2030 and those stressing a continued growth to 2030, ultimately rest upon different assumptions for URR (UKERC 2009).
6. For the last decade, approximately 65 per cent of new booked reserves come from new estimates of existing reserves in old fields. The amount of reserves is usually highly underestimated at the beginning of the development process.
7. Once again the definition of non-conventional oil is a subject of debate, coal-to-liquid and gas-to-liquid being included or not. Therefore, a comparative perspective is not straightforward.
8. Statement of David P. Russ Regional, US Geological Survey, Before the Senate Energy and Natural Resources Committee Water and Power Subcommittee to

Examine Shale Gas Production and Water Resources in the Eastern United States, 20 October 2011.

9. Directorate General for Internal Policies, *Impacts of Shale Gas and Shale Oil Extraction on the Environment and on Human Health*, IP/A/ENVI/ST/2011-07, June 2011.

10. IHS-CERA, Oil Sands, Greenhouse Gases, and US Oil Supply, Getting the Numbers Right, 2010.

11. The only quantitative objective that came out of the fifteenth Conference of the Parties in Copenhagen.

12. Security of energy considering its uncertainty, risk and economic implications, SECURE, project N°213744 under EU DG-Research FP7

http://cordis.europa.eu/search/index.cfm?fuseaction=proj.document&PJ_RCN=9905806

Policy on Natural Resources, POLINARES, project N°244516 under EU DG-Research FP7 [Accessed 26 October 2012].

http://cordis.europa.eu/search/index.cfm?fuseaction=proj.document&PJ_RCN=11275558 [Accessed 26 October 2012].

10
Peak Gas: Technology and Environmental Consequences

Ariel Bergmann

Peak gas is just one more expression of the concern society has over the use and overdependence on a finite resource to sustain economic activities. It is closely tied to the debate over peak oil and peak minerals, as discussed in Chapters 9 and 11. The peak gas thesis proposes that some finite amount of gas is technically recoverable in the world and that, at some point in time, maximum production will peak and will be followed by a decline with resulting higher costs of extraction and shortages. In its most basic form, this must be true. Natural gas supplies are finite and, if humans continue to consume them, they will become exhausted. However, with the advent of new technologies that 'point in time' has shifted further into the future, possibly by hundreds of years, but definitely for decades, given the increased economical viability of the development of unconventional natural gas resources. This chapter will not debate or discuss the validity of the peak gas thesis, but rather acknowledge that it will occur 'someday'. The real question that this chapter will address is the environmental consequences from the use of new technologies to make more natural gas available and what should be our response to the concern over the environmental consequences.

In 2011, the International Energy Agency (IEA) posed the question: 'Are we entering a golden age of gas?' as the title of a special report (IEA-WEO 2011). This report presented information implying that natural gas could fill a significant portion of the growing global demand for energy over the next 25 years and beyond. That report was followed a year later by another report titled, 'Golden Rules for a Golden Age of Gas' (IEA-WEO 2012). In this later report the IEA acknowledged that there was a growing resistance by civil society and stakeholders, both in the US and in Europe, who were expressing objections based on the potentially harmful environmental impacts from shale gas development and the risks posed by hydraulic fracturing.

This chapter focuses primarily on the case study of how shale gas development has proceeded in the US but also includes some comparable information on the European experience and the Chinese potential. The availability of natural gas in North America, and potentially the world, has entered a new era during the last decade. The confluence of several technologies, some new and still evolving rapidly, some mature and well established, has led to this period of increasing abundance. At the same time, new concerns have arisen as to the environmental and human health risks that accompany shale gas projects. This chapter begins by presenting the increased availability of natural gas in the US and offers forecasts of future supplies. This is followed by a discussion and brief history of some of the technologies that have made shale gas extraction economically possible and the environmental risks that are a source of public concern, in particular the potential for contaminating both ground and surface water. Finally, examples of how differing levels of American government and European countries have responded to shale gas development are given, followed with a brief analysis and concluding remarks on these reactions.

Production of natural gas in the US and other selected markets

In the US, the production of unconventional natural gas, shale gas specifically, has experienced significant increases in the past decade and is forecast to dominate growth in overall US natural gas production in the coming 25 years (EIA 2012b). Technology is opening up new possibilities for unconventional gas to play a major role in the future American energy mix, a development that is already abating concerns about reliability, affordability and security of supply of natural gas. The impact within the US has implications for global reliability, affordability and security of supply as well.

Shale gas currently has a high degree of uncertainty in regard to its potential reserves as would be expected with such a relatively new extraction process and limited geological experience. As set out in Figure 10.1, the US Energy Information Agency (EIA) forecasted 83 trillion cubic feet (TCF) of technically recoverable reserves (TRR)[1] for shale gas in 2004 (EIA 2012a). Four years later, in 2008, the TRR estimate had increased by 410 per cent to 347 TCF and then experienced a further doubling between 2008 and 2009 to 827 TCF.[2] However, experience with the geology and the effectiveness of the drilling process led to a re-evaluation of the TRR and the estimated US reserves were lowered by 345 TCF in 2010. Even this lower quantity, 482 TCF, represents a five-fold increase over the 2004 estimate. This volatility in estimated quantities and supply forecasts is expected to diminish as more experience is gained in the major basins.

Shale gas is generally found throughout central US, from the western Appalachian Mountains to east of the Inter-Mountain Region of the American West (EIA 2011b). The Barnett Basin in Texas was the location of early research and development efforts (1980s and 1990s) and became the first major producer of gas, but the Appalachian Basin in the Mid-Atlantic/Northeast may contain the largest TRR (NETL 2011). The Marcellus

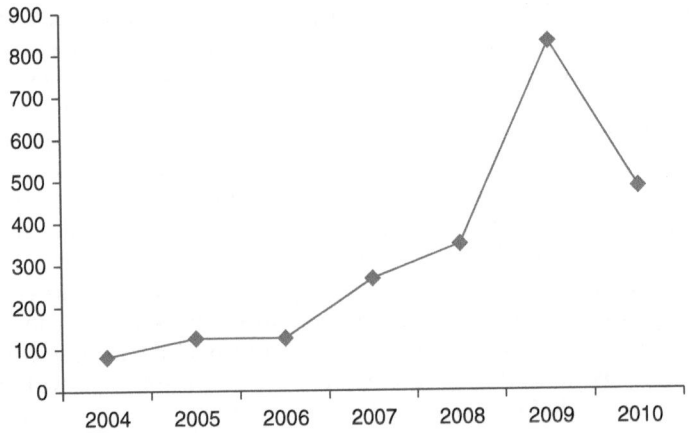

Figure 10.1   Unproven technically recoverable shale gas reserves (trillion cubic feet)
Source: Annual Energy Outlook 2012 (EIA 2012b).

Basin, which is located in nine US states, is currently attracting the most significant interest and development activities. After a long period of steady growth from 2006 to 2010, growth in total US daily dry gas production levelled off during the first three months of 2012, averaging 63.8 BCF per day (23.3 TCF per year), a level almost 9 per cent above the same period in 2011. Production from the Marcellus formation accounted for much of the year-over-year growth in dry gas production.

The reference case presented in the EIA-AEO 2012 forecasts continued natural gas production growth through 2035, the length of its forecast (EIA 2012b). The main source of this growth is ever-increasing development of shale gas resources, as seen in Figure 10.2. The forecast predicts little change in production levels from conventional and offshore gas fields within the US. By 2035 shale gas is projected to be 49 per cent of total US production, more than double its 23 per cent share in 2010. In the reference case previously mentioned, the estimated proved and unproved shale gas reserves will have a combined volume of 542 TCF, out of a total US reserve of 2203 TCF.

Given the increased availability of natural gas and the expectation that domestic supply will exceed domestic demand by the early 2020s, opportunities for gas exports will be created (IEA 2012). In 2010, the US imported 11 per cent of its total natural gas supply, almost entirely by pipeline from Canada and Mexico (EIA 2012f). However, imports from Canada, Mexico and LNG shipments fell significantly in 2011 and 2012 as lower-priced domestic supplies became available. This continued a decline in imports that started in 2007. The average level of LNG imports during the first three months of 2012 were 0.6 BCFD, down about 44 per cent from a comparable period in 2011. LNG imports through US terminals peaked in 2007

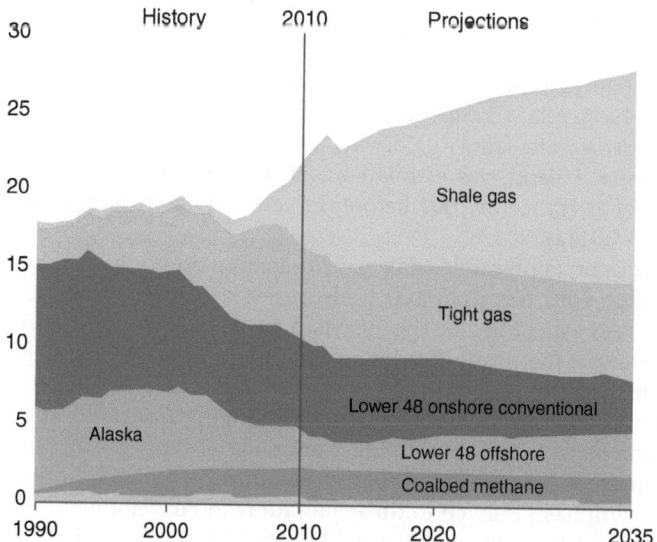

Figure 10.2 Natural gas production by source, 1990–2035
Source: Annual Energy Outlook 2012 (EIA 2012b).

at over 2.1 BCFD. Higher natural gas prices in competing markets abroad are attracting 'spot' LNG cargoes that can be delivered under flexible pricing terms. Currently, only one new export license has been issued by the US government to export natural gas and this will be achieved through the construction of a new export facility adjacent to an existing import facility. The US Department of Energy (DOE) has an obligation to approve LNG export applications in a timely fashion for gas being delivered to countries with which the US has free trade agreements (FTA). Licenses for export to non-FTA countries can only be granted if it is demonstrated to be in the public interest.

Europe and China

Poland appears to be taking the lead in developing shale gas resources so as to reduce dependence on Russian natural gas and possibly export gas to its EU partners. The country has one of the largest shale gas reserve estimates in Europe with a TRR of 12.2 TCF–27.1 TCF. As with the volatile US estimates, the Polish Geological Institute revised their estimate down by over 85 per cent in 2012 from 187 TCF. Commercial production is likely to commence by 2014–16, if the country is able to attract sufficient financial resources and technology from international oil companies, as Poland is deficient in both areas (Natural Gas Europe 2012). An announcement of environmental regulations addressing shale gas developments is predicted for late 2012 and which would take effect in 2013. Other European countries known to have shale gas reserves and who are actively examining development options include Germany, Netherlands, UK, Spain, Romania, Lithuania, Ukraine and Denmark. Additional countries that have known reserves are France, Norway, Sweden, Turkey, Romania, Hungary and Bulgaria. Poland was estimated to have a similar level of TRR shale gas reserves as France in 2009 before the revision downward, as mentioned above (EIA 2011a).

China has no appreciable shale gas production and only a fledgling shale gas industry, with less than 100 wells in test development areas. The first shale gas well was drilled in the Weiyuan gas field (Sichuan basin) in 2010. But there exists the possibility for shale gas to play a major role in China's energy mix, if their announced plans are fulfilled during the current five-year plan. With current reserve estimates ranging from 886 TCF to 1094 TCF, China plans to produce 230 BCF annually of shale gas by 2015, which would represent 2–3 per cent of total gas production in 2015. During this initial phase emphasis is given to the exploration and development of reserves. Slightly longer-term plans call for 2119 BCF (2.1TCF) of shale gas to be produced in 2020 as large-scale commercialization occurs (Platts 2012).

Types of natural gas reserves

Conventional natural gas is generally described as having two key characteristics (Natural Resource Canada 2011). The first characteristic describes its association with oil. It is called 'associated gas' if it is either dissolved in the oil or is found as a cap of gas above the oil. It is called 'non-associated gas' if it is found in reservoirs that do not contain significant crude oil or where the supply of oil is small and the production of the gas does not appreciably affect recovery of the crude oil. The second characteristic is if it is located in conventional geologic formations and is extractable with established cost-efficient technologies.

There is no widely acknowledged consensus on what constitutes unconventional gas at this time. As can be seen in a statement by the EIA:

> what has qualified as 'unconventional' at any particular time is a complex interactive function of resource characteristics, the available exploration and production technologies, the current economic environment, and the scale, frequency, and duration of production from the resource. Perceptions of these factors inevitably change over time and they often differ among users of the term.
>
> (EIA 2012a)

Given this caveat, there are six main categories of unconventional natural gas (NaturalGas.org 2012): deep gas, tight gas, shale gas, coal-bed methane,

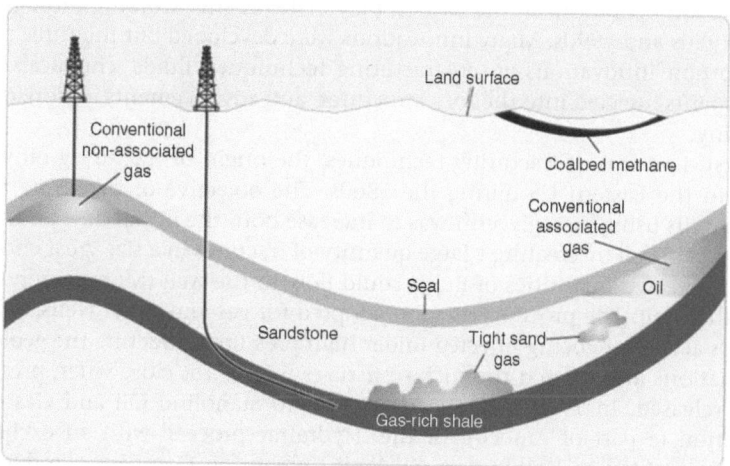

Figure 10.3 Geology of natural gas resources
Source: EIA Energy in Brief. http://www.eia.gov/energy_in_brief/about_shale_gas.cfm.

geopressurized zones and sub-sea methane hydrates (see Figure 10.3). Deep natural gas is typically found at depths greater than 15,000 feet in conventional geologic formations. Tight gas is contained in unusually impermeable and non-porous rock, sandstone or limestone. Coal seams and surrounding rocks may also contain commercial quantities of gas. Geopressurized zones are underground formations that are under unusually high pressure at depths between 10,000 and 25,000 feet below the surface. Methane hydrates are formations made up of a lattice of frozen water, which forms a sort of 'cage' around molecules of methane and were first discovered in permafrost regions of the Arctic. The US Geological Survey estimates that methane hydrates may contain more organic carbon than the world's combined total of coal, oil and conventional natural gas. Shale gas is found in shale formations containing significant accumulations of natural gas.

History of technology research and development

The increased supply of shale gas in the US during the past decade is the result of a confluence of several technologies developed in the oil and gas industry. These technological developments resulted from decades of private-industry investment in research and innovation and government-funded research into technologies that showed potential to increase available reserves of domestic oil and gas resources. Mitchell Oil is credited with being the pioneering company that first developed economically viable gas extraction operations in the Barnett play of Texas after working throughout the 1980s and 1990s. Many innovations were developed but the three most important innovations are in fracturing techniques; fluids, chemicals and proppants injected into the rock structures; and advancements in horizontal drilling.

First, in terms of fracturing techniques, the origin of fracturing oil wells was in the Eastern US during the 1860s. The objective of fracturing hard rock wells using nitroglycerin was to increase both the initial flow and total recoverable oil by creating a large quantity of fractures in a size great enough that increased quantities of fluids could flow to the well (Montgomery and Smith 2010). The process was soon adopted for gas and water wells. By the 1930s acids were being injected under high pressure to fracture the geologic formations and to etch the rock so fissures would not close when pressure was released. In 1949 a patent was issued to Stanolind Oil and Gas Corporation (a part of Amoco) for the 'Hydrafrac process' with an exclusive license granted to Halliburton Oil Well Cementing Company. Production of oil wells increased an average of 75 per cent for the first 332 wells that were hydrafactured that same year. Treatments, as a single iteration of the fracturing process is known, on average consist of 750 gallons of fluid (a gelled crude oil) and 400 pounds of sand injected under several thousand pounds of pressure. The sand is partially retained in the rock structure and acts to hold prop open the fractures. By 2010, treatments averaged 60,000 gallons of fluid and 100,000 pounds of propping agents (sand or sand substitutes), while large treatments have surpassed 1 million gallons of fluid and 5 million pounds of proppant. The procedure is now used by the global oil and gas industry with 50,000 fracture treatments performed in 2008, with the majority being applied in shale gas or shale oil deposits (Montgomery and Smith 2010).

For the second set of innovations, the first liquid used in 1949 was gelled crude oil, but fiscal and technical requirements have required a move to other fluids (Montgomery and Smith 2010). Fracturing fluids need to have low viscosities which create less friction during treatment and so that lower pressures are required during injection, but they also have to be sufficiently viscous to suspend and carry the proppant or sand as well as be inexpensive. Water was first used in fracturing treatments in 1953. At that time, and in the following decades, chemicals were added to the water to increase the viscosity, stabilize the solution, and minimize the effect of water on certain types of geologic formations. Approximately 96 per cent of treatments that inject proppants using fracturing fluids are based on water, brine or acids as the main fluid (Montgomery and Smith 2010). The treatment of higher-temperature wells requires additional additives such as metal-based agents and methanol to stabilize the gelled water. Further chemicals and agents have been developed to allow for even greater usage of fracturing in more challenging geology. Many of the additives used in the process are harmful or toxic to humans if consumed and thus pose an environmental risk if surface or ground water is contaminated.

River sand was the first material used as a proppant, followed by construction-grade sand of a uniform size. Several alternative propping agents have been evaluated in the last 60 years, such as glass beads, steel shot, aluminium and plastic pellets, resin-coated sands and others (Montgomery and Smith 2010). The amount of sand to fluid was a relatively low ratio until the viscosity of fracturing fluids could be chemically controlled to allow for increased proppant size. The improvement in pumping equipment has also facilitated higher concentrations of proppant to fluid levels.

For the third key innovation, horizontal drilling, the first recorded true horizontal oil well was completed in Texas in 1929. A second was drilled in 15 years later in Pennsylvania, to a depth of 500 feet (Helms 2008). Practical application of the technology did not materialize until the early 1980s, when the downhole drilling motors were sufficiently developed and downhole telemetry equipment was invented. These two technologies combined to make horizontal drilling technology efficient and commercially viable. The second generation of horizontal drilling during the 1980s and 1990s resulted in horizontal lengths growing from 400 feet to 8000 feet (Helms 2008).The third and current generation of horizontal drilling has resulted in much longer, deeper and more accurate placement of multiple horizontal well bores that can exploit the relatively narrow gas bearing strata of shale that contains natural gas (Helms 2008).

Environmental consequences

According to the Energy Institute at the University of Texas there are environmental risks associated with all phases of shale gas development and production, as with all oil and gas projects (Energy Institute 2012). Many actors in civil society are strongly focused on the potentially adverse environmental impacts of hydraulic fracturing (Bloomberg 2012). Eight categories of operational or environmental risk and impact issues were identified in the Energy Institute's report but only the water issues and atmospheric emissions are going to be addressed in this chapter.

The most significant environmental contamination risk from shale gas projects to local water resources occur during the drilling phases. There are three specific issues and actions that have created public anxiety over shale gas operations. The first, and seemingly the most publically exposed, is the possibility of fracturing fluids contaminating underground water resources. The second is how the fracturing fluid, after use in the fracturing process, is managed once it is withdrawn from the well. And third is the quantity of water required for fracturing treatments and the impact it has on local water availability and other water users.

Less than 1 per cent of the fracturing fluid consists of chemical additives; the balance is approximately 90 per cent water and 9 per cent proppants. Some of the chemical additives incorporated in the fluid are toxic or carcinogenic, according a Congressional report (US House of Representatives 2011), and therefore cannot be released into the open environment with the potential for harm to humans and the ecosystem. Specifically, the concern is that fracture fluids may escape during the drilling process and migrate into underground aquifers which may be used by the public for household consumption, agriculture or commerce. Once the fracturing process has been completed, the well is depressurized and the injected fluid returns to the surface along with saline water from within the shale deposit. The volume of water returned to the well head after depressurization, compared to the amount of injected fluid, can range from three times as much returned to less than one-sixth the amount.

This concern is not unfounded as the US Environmental Protection Agency (EPA) released a preliminary report in late 2011 that stated that an aquifer in Wyoming showed chemical contamination that likely came from fracturing activities in the area. Synthetic chemicals, like glycols and alcohols consistent with gas production and hydraulic fracturing fluids, benzene and high methane levels were identified. They further stated, 'the presence of these compounds is consistent with migration from areas of gas production' (EPA 2011). Substantial quantities of water are required for shale gas development. With increasingly longer horizontal wells the average quantity of water used by each well has increased from four to six million US gallons being used in the most active basins like Barnett and Marcellus. With thousands of wells being drilled and millions of gallons of water being used by this new water consumer, some communities with limited water resources are facing a dilemma over the provision of treated water and the allocation of water from surface or ground sources.

A report commissioned by the European Commission and published by the AEA (2012) examined the potential risks for the environment and human health from hydraulic fracturing operations in Europe. This report accessed a limited number of categories to be high risk when individual drilling sites were considered. However, numerous categories were accessed as high risk when the cumulative number of projects was taken into consideration; high risk was found for ground- and surface-water contamination, adverse impact on water resources, air release of pollutants, land take, biodiversity, noise impacts and traffic. The only categories found to be non-high risk were the visual impact at moderate risk and seismicity at low risk.

One positive outcome from expanded supply of natural gas in the US and the resulting decline in market price is the increased use of gas by switching from coal for electric power generation. CO_2 emissions from coal were down 18 per cent in the January–March 2012 period. That is the lowest first-quarter CO_2 emissions from coal since 1983 and the lowest for any quarter since second quarter of 1986. The decline in coal-related emissions is due mainly to utilities using less coal for electricity generation as they burnt more low-priced natural gas (EIA 2012g). Natural gas emits lower amounts of CO_2 in comparison to these other fossil fuels. Nonetheless, an increased share of natural gas in the global energy mix is not sufficient on its own to put the world on a carbon emissions path consistent with an average global temperature rise of no more than 2°C (IEA 2012).

Responses to these environmental challenges

In the US, responses to the environmental concerns expressed by civil society and local stakeholders from shale gas development have occurred at three levels. Global institutions, like the International Energy Agency, have published evaluations of this potential resource, but more importantly issued guiding advice on how governments (local, state and federal level), businesses and other stakeholders can engage in a constructive manner to diminish the risks from development and implement good business practices for the gas industry. The US federal government has engaged both the legislative and executive branches in providing legislation and regulatory oversight. State, county and local municipal governments are the most active, in some ways, of all the institutions engaged with the industry and stakeholders. This level of engagement is the result of environmental impacts occurring primarily at a local/regional scale, especially for water issues.

In response to the environmental concerns being expressed, several US states and local governments have passed bans and moratoria on horizontal drilling and hydraulic fracturing activities while reviews have been or are currently being conducted, so that appropriate new regulations could or can be established. New York State enacted a drilling moratorium in 2008 when the NY Department of Environmental Conservation (DEC) began an environmental review of horizontal drilling and high-volume hydraulic fracturing. More than 30 municipalities in upstate New York have passed bans on gas drilling and more than 80 have enacted moratoriums in anticipation of DEC completing its environmental review and lifting the four-year-old state moratorium. Two New York judges recently upheld local ordinances banning the practice of moratoriums, whereas one judge in West Virginia ruled a local ordinance unconstitutional and unenforceable (RFF 2012). New Jersey and Maryland both enacted state-wide moratoriums following New York's lead. In 2012, North Carolina passed legislation allowing horizontal drilling in principle, but drilling will not be allowed until a regulatory framework is in place, which should occur by 2014 (RFF 2012).

Various states have enacted rules and regulations for shale gas development (RFF 2012). Examples include, first, pre-drilling stage water-well testing. Pre-drilling tests of well water to establish the baseline water quality for an area prior to drilling activities is mandated in some states. American Petroleum Institute's (API) best practice recommendation is to test samples from any source of water located near the well (distance is based on anticipated fracture lengths) before drilling or hydraulic fracturing starts (API 2010). Second, in terms of water withdrawals, most states require general permits for surface water and/or groundwater withdrawals. However, some states have a minimum volume requirement below which permits are not required. Pennsylvania requires a water-management plan covering the full life cycle of the water used in shale gas production, including the location and amount of the withdrawal and an analysis of the impact of the withdrawal on the body of water from which it came. Kentucky specifically exempts the oil and gas industry from water withdrawal requirements. API best practices state that 'consultation with appropriate water management agencies is a must', and 'whenever practicable operators should consider using non-potable water for drilling and hydraulic fracturing' (API 2010a and 2011).

A third example of state-enacted rules and legislation is those for casing and cementing depth. Some states have set the required depth for casing and cementing the well to a predetermined depth below the water table. This is done to prevent the well from leaking and thus protect the underground water table. API best practice is a casing and cementing depth of 100 feet below the deepest underground source of drinking water encountered while drilling the well (API 2009 and 2010b).

In addition, almost all of the states surveyed allow deep well injection as a form of wastewater disposal. North Carolina prohibits underground injection of fluids produced in the extraction of oil and gas wells. Arkansas has a moratorium on deep injection in a 600-square-mile area of the state where a fault may have been activated by wastewater injections in the area. Ohio recently followed a similar course of action, temporarily closing down several injection wells in an area where seismic activity has occurred. Arkansas, Colorado, Ohio, Oklahoma and Texas have recently experienced an increase in seismic activity near deep underground injection sites. API best practice is that 'disposal of flowback fluids through injection, where an injection zone is available, is widely recognized as being environmentally sound, is well regulated, and has been proven effective' (API 2010a and 2011).

Responses at the federal level

The EPA is the key federal agency in regards to shale gas development within the US. The environmental concerns that are being voiced in the US have to be addressed by this agency. The EPA's focus and obligations under the law are to provide oversight, guidance and appropriate regulation to achieve adequate protections for the air, water and land within the US (EPA 2012a). The EPA is working with states and other key stakeholders to ensure that shale gas extraction does not degrade the environment or endanger public health. The agency engages and addresses hydraulic fracturing on several fronts (EPA 2012a). It works to improve the scientific understanding of hydraulic fracturing, provide regulatory clarity with respect to existing laws, and, using existing authorities where appropriate, to enhance health and environmental safeguards. Currently the EPA is undertaking a national study to determine potential impacts of hydraulic fracturing on drinking water resources. Since the 1970s, the EPA has been given competency through the Safe Water Drinking Act (SWDA) to oversee and regulate underground injection wells by setting requirements for proper well siting, construction and operation to minimize risks to underground sources of drinking water (EPA 2012b). However, the Energy Policy Act of 2005 (US Government 2005) excluded hydraulic fracturing from regulation under the SWDA programme, if the fracturing is for the extraction of oil, gas or geothermal heat production. Regulation is only allowed when diesel fuel is used in the fracturing fluid. EPA has developed a draft permitting guidance document specific to oil and gas hydraulic fracturing activities using diesel fuels. EPA and states share primary responsibility for implementing this programme.

The EPA is investigating and reviewing the different disposal methods employed by the shale fracturing industry to ensure that there are adequate regulatory and permitting frameworks to provide safe and legal options for disposal of flowback and produced water. Underground injection to dispose of fluids or other substances from shale gas extraction is a common method in many regions of the US. Disposal of flowback and produced water by underground injection is fully regulated under the Safe Drinking Water Act (EPA 2012a). The Clean Water Act (CWA) (EPA 2012c) enables the EPA to establish guidelines to regulate and prohibit the on-site direct discharge of wastewater from shale gas extraction into waters of the US. Even though a portion of wastewater from shale gas projects is recycled in to the well, a significant amount still requires disposal. No national standards exist for the disposal of wastewater discharged from shale gas extraction activities. A consequence of this is that some shale gas wastewater is disposed at wastewater-treatment plants creating an environmental risk as not all of these plants are equipped to treat this type of wastewater. The EPA has commenced a process to develop standards for treatment of wastewater produced during shale gas extraction. Regulations are anticipated by 2014.

The EPA, along with other federal departments, agencies and state governments are studying the air pollution emissions that are impacting areas with hydraulic fracturing activities. Increased emissions of methane, volatile organic compounds (VOCs) and hazardous air pollutants (HAPs) have been well documented when shale gas development has occurred in regions of the US (EPA 2012d). In conjunction with industry, cost-effective technologies and proven drilling practices have been identified that can reduce methane emissions that currently escapes to the air from shale gas development and operation in the US. Proposed new rules would reduce VOCs emissions by approximately 25 per cent across the national oil and gas industry. Specifically, it would lead to an estimated reduction of 95 per cent in VOCs emitted from new and modified hydraulically fractured gas wells (EPA 2012e).

European responses

Currently no EU-level directives or regulations specifically governing shale gas projects exist as they do for comparable forms of fossil fuel projects. However, there are two directives, the Water Framework Directive and the Mining Waste Directive, to which shale gas projects have to comply. No EU legal obligation exists that requires project developers to provide Environmental Impact Assessments (AEA 2012). The AEA report outlined a series of options for the EU to improve the regulatory framework, ranging from extending existing regulations to applying them in a different way to introducing new regulations specific to shale gas extraction. There are gaps that could leave the sector under-regulated if action is not taken.

However, member states are determining their own regulatory response, such as France which has imposed a moratorium on shale gas projects. France's President Francois Hollande has stated that no shale gas development will be allowed during his time in office, citing the heavy risk to health and the environment as reason. At the start of 2012, Bulgaria reversed a decision to proceed with shale gas licensing and implemented a moratorium on shale drilling. Romania has also put shale gas exploration on hold, and the Czech Republic is considering a similar move.

Responses at the global level

The International Energy Agency in a *World Energy Outlook Special Report on Unconventional Gas* identified numerous principles that can facilitate governments, industry and civil society stakeholders in addressing the environmental and social impacts of the shale gas development with the purpose of increasing acceptance and minimizing opposition (IEA-WEO 2012). These rules are essentially sustainable business practices that can enable shale gas developers to reduce conflict with local and regional communities and behave in a way that a social license to operate is granted by civil society. It is important for shale gas developers to act in a manner that a social license to operate in communities is established in order to reduce delays or disruptions to operations. Social and environmental responsibilities should be broadly recognized and openly addressed by engagement with communities and stakeholders in each phase of project development. Engagement should start prior to exploration activities and provide adequate opportunity for comment on plans, operations and performance. Developers should take note of concerns and respond in a manner that acknowledges the concerns being raised. Developers need to ensure that some economic benefits are delivered as widely as possible to the local communities.

Baseline indicators for key environmental characteristics should be established prior to commencing any activity with environmental monitoring continuing during operations. Data on water usage, wastewater volumes and contaminants, air pollution emissions including methane, as well as full disclosure of fracturing fluid additives and volumes even when it is not mandatory needs to disclosed in a transparent manner and available to all stakeholders. Developers need to recognize the case for independent evaluation and verification of environmental performance. Drilling sites should be selected to minimize impacts on the surrounding community. Considerations such as existing land use, local work opportunities, heritage and culture and ecology need to be included. Adequate geologic surveying should be conducted to properly assess the risk of earthquakes from interaction with deep faults or other geological features and to avoid fluids passing between geological strata with operational monitoring in place so as to ensure that hydraulic fracturing only occurs in gas producing shale formations.

Governments, at all levels, should initiate appropriate regulatory regimes that are supported by proportionate resources, including sufficient permitting and compliance staff, that reflect the level of development activity and economic value that shale gas may deliver. There needs to be strong and broad-based political support for regulatory regimes. Regulation should seek a balance in policymaking between prescriptive regulation and performance-based regulation in order to guarantee high operational standards while also promoting innovation and technological improvement. Regulators need to be adaptable and pursue continuous improvement of regulations as experience and information on the impacts of hydraulic fracturing drilling operating practices accumulates. Rules and regulations should be implemented for gas field operations that isolate and adequately protect freshwater aquifers and other geologic strata from contamination by drilling, fracturing and extraction activities. Minimum depth requirements should be implemented for hydraulic fracturing to assure public confidence that the project is a safe distance below ground water. In addition, environmental control procedures need to be in place to prevent and contain surface spills and leaks from wells and to ensure proper disposal of waste fluids and solids. Produced and wastewater must be stored and disposed of safely. Both developers and local governments must ensure that emergency response plans are robust and match the scale of risk.

Air emissions should be minimized by designing projects to have zero venting and negligible flaring of methane during well completion. Operations should be conducted to reduce fugitive and vented greenhouse-gas emissions during the entire productive life of a well. Air pollution from vehicles, drilling rig engines, pump engines and compressors should also be minimized. Operations should also be conducted so as to minimize freshwater use and lessen the demand on local water resources. Recycling of fracturing fluids should be used whenever practicable. Environmentally benign fracturing fluids should be developed and promoted to minimize the use of chemical additives. Finally, project and gas field developers should pursue coordinated development of local infrastructure and opportunities for economies of scale that can reduce environmental impacts on the region. Consideration for the cumulative effects of numerous drilling and production activities on the environment quality of a community should be included in operation decisions.

Analysis and conclusion

The institutions of the European Union and the US are competently addressing and balancing the economic potential and need for abundant natural gas and the environmental risks that are posed by shale gas extraction. The US has a mature robust democratic political system at all levels of government, ranging from municipalities and counties to the state and federal levels. The political institutions are generally seen as being inclusive, equitable, following a democratic process of dialogue and participation in the legislative and regulatory or rule-making process. They are held accountable to the rule of law by independent judiciaries. The same can be said for the European Union and its individual member states. The same cannot be said for China, which continues to demonstrate significant limitations on public engagement on economic development issues, and a lack of transparency on environmental regulation and enforcement.

The European Union and the US are on similar paths in developing their shale gas resources, with the major difference being the US is several years ahead in actual deployment of extraction projects. There are other differences that may account for the different speed of advancement; environmental values and relative impact from development, level of state control and ownership of the resource, different geologic structures and depths of the shale strata, costs of extraction, technologic competency, energy security concerns, the role of natural gas in the economy, alternative natural gas sources.

In the US and Europe two important areas have been identified as lacking adequate institutional practice up to this date: (1) the states' capacity to regulate and (2) transparency of the industry on substantial environmental risks to the general public. The shale gas industry has expanded so quickly across America that few states and local governments had any knowledge, let alone capacity, to anticipate the impacts, both positive and negative, that have come upon them. Many states have been playing a catch-up game for several years and are just starting to put into place the capacity to address the local impacts that come from shale gas projects. This situation is best exemplified by the bans and moratoriums on shale gas development that have been enacted in several states. Europe is no better on this account, but is benefiting from lagging behind the American experience. By observing the industry practice and response of civil society in the US, the European Commission and member states are having an opportunity to collect adequate information to deliberate and establish proper regulation before shale gas projects are being deployed.

The lack of transparency by industry stakeholders over potentially dangerous or toxic chemicals used in the drilling and fracturing process have been the key source of public concerns and opposition to shale gas operations. While several governments now require public disclosure of the environmental and health risks, many still do not, and industry continues to oppose disclosure of such information based on proprietary rights. The issue of transparency has the potential to severally constrict any further expansion of shale gas projects in the US and even lead to the prohibition of this technology in Europe and aborting the industry.

Notes

1. Technically recoverable reserves are estimated gas deposits that can be developed using current recovery technology but without reference to economic profitability.
2. For comparison, the US averaged 22.4 TCF total annual consumption of natural gas from 2000 to 2006 (EIA 2012b).

11
Mineral Depletion and Peak Production

Magnus Ericsson and Patrik Söderholm

Natural resources are critical for the economic development of human societies and cultures, and fears of an impending depletion of these resources have been expressed (at least) since antiquity (e.g. Maurice and Smithson 1984). The most recent – and overall very influential – predictions of resource depletion are those concerning the production of oil. Advocates of the so-called peak oil concept suggest that oil production is close to an unavoidable (geologically determined) peak that could have serious consequences for the global economy and society as a whole.[1] The 'peak' concept has increasingly influenced the debate over mineral depletion as well, and some analysts claim that world production of many minerals (e.g. lead, mercury, cadmium) has already peaked or are close to peaking (Bardi and Pagani 2007).

In this chapter we provide a brief economic critique of the 'peak' concept as it applies to understanding mineral depletion. It should be noted that the peak resource hypotheses come in many different shapes (see below). In this chapter we focus on what Helm (2011) refers to as the 'supply-side' peak hypothesis, essentially stating that the earth's physical supplies of specific resources (e.g. oil, copper etc.) are well researched and cannot therefore be expected to support production beyond a certain limit. Despite the emphasis on geological constraints, the peak debate is not concerned with the issue of whether oil or mineral resources will deplete in a physical sense. Long before the last ounce of metal is extracted from the earth's crust, costs would rise, at first curtailing but eventually completely eliminating demand. In other words, what we could fear is not physical depletion, where we literally run out of mineral resources, but economic depletion, where the costs of producing and using mineral commodities increase to the point where no one is willing to buy them. With some qualification (see below) this view appears to be shared by both economists and peak modellers.

The controversy lies rather in how one should model and interpret changes in mineral production and deposit discovery over time. The peak models (e.g. the so-called Hubbert curve) essentially assume that geology is the prime constraint of discovery and mineral production, and they thus only leave room for economic and political factors as *ex post* explanations. This is a serious weakness, and for minerals it may imply that these methods tend to systematically (and often grossly) underestimate the time left before mineral production peaks (also assuming that any anticipated peak will be the final one). This does not preclude the fact that there may come a day when the production of some minerals peaks; in some instances global metal production may peak early not as a result of depletion but, for instance, due to environmental concerns or technological change. Notably, asbestos and mercury experienced their (so far) only peaks in the 1970s as their detrimental health effects became known. Thus, unless we understand the real reasons behind such peaks, decision-makers at the corporate and/or policy levels will be little helped.

In the following section we briefly discuss the fundamentals of the supply-side 'peak' approach, and highlight a number of critical assumptions underlying the methods used. The third section presents a brief economic critique of the peak approach and suggests a closer synthesis of geological and economic knowledge. In the fourth section we discuss whether the peak approach is more suitable for the case of oil compared to non-energy minerals. Some final remarks are then outlined.

The peak approach to mineral depletion

The Hubbert peak and more recent approaches

As was noted above, there are several peak resource hypotheses, including claims about the physical geography of reserves, the recovery rates and the price of mineral resources. In this chapter we address primarily the peak approaches that build on the work by the oil geologist M. M. King Hubbert in the 1950s (Hubbert 1956, 1971). The model used by Hubbert was based on logistic growth curves (bell-shaped curves), and if applied to minerals it essentially suggests that geology requires that the production of the mineral should follow such a curve and thus peak at a stage where (roughly) half of the ultimately recoverable resources (URR) have been extracted. URR is assumed to be known, and it refers to the amount of the mineral that is believed to be recoverable given existing technology.[2] URR may include estimates of undiscovered resources, although normally this represents only a fraction of the total. Still, a critical assumption is that the technology of exploration and exploitation of resources is now so mature that believable ultimate quantities can be determined. In practice estimates of URR are very rarely static, and instead change due to the influence of political and economic factors (Pesaran and Samiei 1995; Radetzki 2010).

The additional assumptions underlying Hubbert's 'peak curve' include also that (a) the population of producing deposits is sufficiently large so that the sum of all fields approaches a normal distribution; (b) the largest deposits are discovered and developed first; and (c) mineral production continues at its maximum possible rate over time (Bentley 2002). All these assumptions may well be seriously criticized (Lynch 2003), and we revert to such critique in next section. Hubbert used this approach to predict that US oil production would peak in the early 1970s, which in fact it did.

During recent decades Hubbert's approach has been extended to the global scale, including also methodological developments. For instance, while the original Hubbert curve was defined as a bell-shaped curve, more recent approaches have employed curves that are mathematically represented using the derivative of the logistic function and also generated using Gaussian function (May et al. 2011). In the oil case, so-called creaming curves have been employed, describing the relationship between the exploration activity and the total discoveries for a given region (see Chapter 9, this volume). This approach can – if complemented by information about ultimate resource limits – provide a dynamic analytical framework, which also addresses the role of technological progress in oil recovery.

The central message from this type of models is that resources have to be found before they can be produced, implying, for instance, that a drop in discoveries signal increased resource scarcity and as a result a decline in production. Aleklett and Campbell note, for instance, that 'extrapolating the discovery trend of the past to determine future discovery and production should be straightforward' (2003: 5). Production mimics discovery with a certain time lag, because what is produced needs to be discovered first. This notion, it is argued, appears to be independent of the kind of resource considered. In the oil case, these analysts assert, we have – over the recent decades – witnessed a situation where new discoveries have typically been smaller than extraction. For this reason, it is argued, global oil production is bound to peak in 2020 at the latest (see Chapter 9, this volume).[3]

Applications to minerals and metals

In recent years the peak approach has been increasingly applied to minerals as well. Bardi and Pagani (2007) examine the world production of 57 minerals, and conclude that there is 'evidence' that 11 out of these have clearly peaked and production levels are thus declining. Several additional minerals may be peaking or appear to be close to peaking. The authors express strong support for the Hubbert model, and claim that this model captures a nearly universal phenomenon in resource exploitation. This view is supported by Laherrére (2009), who asserts that the historical gold-production curve can be interpreted in terms of multiple production cycles, each one following the Hubbert curve. Giurco et al. (2010) investigate the issues of mineral resource extraction and depletion in Australia, and conclude that the peak minerals concept is useful for addressing the gradual depletion of minerals in the country. The authors conclude, for instance, that copper fits the Hubbert model well and appears to be approaching a peak in the country. Similar to the findings of Laherrére (2009), Giurco et al. (2010) argue that gold production in Australia reflects multiple peaks in production due to technological transitions that have occurred in the industry. The studies on individual countries have, though, very limited interest in an age where markets and economic development are becoming increasingly global. The outlook when securing the future supply of natural resources has to be global, otherwise sub-optimal solutions will be chosen.

The studies employing the Hubbert-type models do not necessarily ignore economic factors (e.g. Bardi and Pagani 2007); they do recognize that peaking and decline in production is the result of gradual increases in the cost of production of the mineral as it gradually depletes over time. These costs can be expressed in monetary terms but they are also sometimes expressed in energy units instead (Laherrére 2009), the latter approach thus emphasizing only one cost component of the mineral-extraction process. However, most peak studies only highlight economic impacts at the supply-side of the minerals market, and, as was noted above, the economic explanations are only provided *ex post* as rationalizations of the projected peaks rather than explicitly integrated into the models.

Thus, Hubbert models ignore the fact that the rate-of-return on mineral investment (in production capacity as well as exploration) will depend on the price of the mineral (which in turn will be influenced by the marginal cost of extraction). Moreover, extraction costs are not only determined by physical depletion but also by demand patterns, which in turn will be affected by the price level and the ease with which consumers can substitute to other minerals, products or services to satisfy human needs. This is a general concern with the peak approach, and this is elaborated on in more detail in the next section. Following that, we revert to the question of whether the peak approach may be judged as more relevant to the case of oil compared to minerals due to the characteristics of the resources and of the demand for oil.

The economic critique of the peak minerals approach

Clearly mineral discovery is a necessary condition for mineral production but it is certainly not a sufficient condition, and although there may exist a (lagged) statistical relationship between the two variables this does not indicate a causal relationship. Data cannot speak for itself without the help of theoretical tools, and economic theory provides some of the most important tools (but certainly not the only one) in this particular area. Introductory economics textbooks teach us that in the presence of growing economic scarcities (e.g. lower ore-grade deposits at more remote locations) the price of minerals will increase inducing: (a) a decrease in global demand; (b) investments in new capacity and exploration; and, not least (c) increased efforts to improve current production and exploration technologies. The outcome of this process is far from certain (it is simply unknown), and can of course be constrained by geological factors and does not preclude the emergence of a peak production. However, if we do not model this process in a consistent manner (addressing the relevant uncertainties) we will learn very little about when this peak may emerge. No one has stated this more aptly than Popper (1957): 'The total supply of any mineral is unknown and unknowable because the future knowledge that would create mineral resource cannot be known before its time.'

Clearly, political factors also play an important role here. In the peak oil debate it is frequently pointed out that depletion is occurring since oil production has systematically exceeded 'new discoveries' since the 1980s. While this is true in itself it does sometimes ignore the very substantial appreciation of oil wells after discovery. The quantity of reserves in new discoveries regularly appreciates in the process of field development exploration and subsequent exploitation. It may often be further appreciated as more efficient technology is taken to use. A similar pattern is found for mineral deposits; the ultimate extraction from a deposit typically represents a quantity several times greater than the initially recorded discovery. Again, the presence of appreciation, even in those cases where attempts are made to control for it, are inherently uncertain.

Moreover, the idea that this relative lack of discoveries must (by some immutable natural law) imply lower production ignores the fact that in the case of oil the drop in discoveries was largely a result of a drop in exploration activities in the Middle East. Governments nationalized foreign operations and exploration expenditures were reduced as demand for oil fell drastically (Lynch 2003). From an economic perspective this makes sense, since it is uneconomical to spend money exploring for something that consumers have become more reluctant to buy. This does not preclude geological factors playing a role in explaining the drop in discovery, but unless we properly address the underlying economic decisions little will be learnt that can inform various types of decision-makers. Similarly, declining demand growth for some metals is overall a result of greater material efficiency and substitution, and not necessarily scarcity. Our concern is thus that the peak approach biases interpretations of market developments towards highlighting (solely) the importance of geological constraints. It should also be noted that in the oil and minerals industries the process of developing new productive capacity is often a lengthy process (even in the absence of fundamental geological constraints), due to environmental permitting requirements, public opinion etc. Thus, following a boom in the demand for minerals the price of minerals may be high for an extended period of time.

The examples given above of previous peaks in the production of lead, mercury and cadmium (Bardi and Pagani 2007) all illustrate how important demand is. For each of these metals a sharp fall in demand, rather than geological availability, resulted in peaks in production. All these three metals have serious detrimental effects both on the environment and directly on man when producing and using them. As these health and environmental effects have become known the metals have been totally or partially banned by government authorities. It is also no doubt possible that the use of one metal could 'peak' for other political reasons. Uranium is one example; after the end of the Cold War the use of weapon-grade uranium made the production of uranium from geological resources drop, which again had nothing to do with the geological resources depleting.

We assert that in contrast to the deterministic approach offered by the Hubbert modellers it needs to be acknowledged that in the long run the seriousness of depletion trends depends 'on the race between the cost-increasing effects of depletion and the cost-reducing effects of new technology. The outcome will be influenced by many factors, and is simply unknown' (Tilton 2003: 41). Historically the real prices of most non-renewable resources point to a declining rather than an increasing long-run trend. This is due to the fact that new discoveries are made (again, as a result of conscious investment and exploration activities), and that technical progress has refined the methods used to find new deposits and lowered the extraction costs and has also made it possible to use substitutes for virgin natural resources (Radetzki 2002). Moreover, the mining companies themselves make reserve estimates, and they typically identify reserves within a fixed time horizon in order to minimize the costs associated with gathering such information (May et al. 2011).

Over the twentieth century the growth in metal production has, in spite of falling prices, been possible because of technological developments. The productivity of, for example, drilling, one of the processes involved in mining, has increased more than one hundred times in the same period, thus overcoming falling ore grades in reserves and the deterioration of other cost-increasing factors. This shows that in many cases the cost-reducing impacts of technical progress and resource substitution have far overcome the cost-increasing impacts of depletion. There are no indications that this process of technological development will stop at this stage. Over time technical progress has taken place in waves and with the presently increasing prices of many mineral commodities, the incentives to invest in R&D in this area have become stronger. The results of these present efforts will take decades to translate into practice and it is still too early to *a priori* say that they will not succeed.

At present the rate-of-return on mineral exploration R&D is high given the high commodity prices, and there is currently great interest in exploring the possibilities to extract minerals from greater depths or from the seabed. In recent years the traditional geological concepts of metalogenesis over millions of years have been challenged as new frontiers of the earth are being explored. The continuous geological processes leading to the phenomena of 'black smokers' and other examples of ore deposits being formed on the bottom of the sea, open up new sources of metals production. For instance, the first commercial concessions for deep sea mining of black smokers have recently been granted in Papua New Guinea (Ericsson 2008). These concessions illustrate both how novel technologies make previously high-cost deposits available, and also how the static geological assumptions underlying the peak theories are being challenged.

At present we do not know about the outcome of these efforts, but given the important role of technical change in the past, these forces cannot simply be ignored by assuming a constant stock of ultimately recoverable resources. Tilton (2003) notes that at current rates of consumption the copper and iron found in the earth's crust would last 120 million years and 2.5 billion years, respectively. Clearly, only a tiny share of this 'stock' is economically available given present technology, but unless the potential future technological progress is taken into account we cannot expect predictions of future mineral-extraction discoveries and extraction rates to perform well. The peak concept does not help us much in this respect, at least until it acknowledges that technological change is heavily induced by economic and political factors.

Are there differences between oil and metals?

Most of the peak studies so far have focused on the case of oil, and Giurco et al. (2010) note that there exist fundamental differences between oil and many minerals such as the fact of the recyclability of metals. Metals are elements and hence are indestructible. Once they are mined and put into use they do not disappear but are in most cases available for recycling, that is, production can increase but the balance of supply between primary and secondary sources will vary.[4] For example, steel production of the US has continued on a high level based on scrap even though iron ore production has not grown at the same pace.

May et al. (2011) review the use of Hubbert-type approaches to the concept of peak minerals. They note that with respect to minerals there are several factors apart from recyclability which make the use of these models more complicated for minerals than for oil. For instance, the presence of varying ore qualities and quantities imply that completely new technology may have to be developed in order for these ores to be exploited. Moreover, generally there is a lack of reliable discovery data for many metals compared to the information about oil reservoirs. Nevertheless, the authors argue, 'by using a range of estimates of resources and/or reserves, a period of time can be identified which indicates when a peak in minerals production may occur' (ibid.: 23).

However, we contend that the main weakness of the Hubbert approach lies not so much in its applicability to certain minerals or energy sources; it concerns a more fundamental methodological problem of ignoring the nature of the causal effects in the minerals (and oil) markets. This makes the peak concept a poor tool for projecting future mineral-production levels, and, perhaps even more importantly, for providing relevant knowledge that can support future policy decisions and investment plans. In the case of ore concentration this has very similar impacts as does the size of the oil fields. At given technological knowledge it becomes more expensive to extract from both low-grade mineral deposits and small oil fields. In addition, the present move from conventional to unconventional oil (see Chapter 9, this volume) is similar to the transfer from selective to mass mining of copper in the early twentieth century (Radetzki 2009). In the case of selective mining metal-rich veins were carefully extracted, while mass mining has involved the full utilization of large low-grade ore bodies.

It is sometimes argued that oil demand differs from that of mineral demand, oil having fewer substitutes. But overall this is also not supported empirically. In the short run the own-price elasticity of demand is very low for both oil and other minerals. In the long run there is considerably greater scope for substitution in both cases, as well as for the introduction of new resource-saving technology.

Final comments

In this review we do not argue that the notion of a peak in minerals production in itself is a foolish one; instead we have criticized the methodological approach used by most peak modellers. We argue that it ignores the underlying causal relationships in the supply and demand for mineral commodities and the associated exploration activities. Clearly minerals are nothing but non-renewable in the sense they cannot be biologically reproduced but this is not of prime interest as they are not – like oil – destroyed when used.

Economic depletion of minerals may become a real concern for mankind in the distant future. Still, the Hubbert curve and any associated approaches do not provide us with much guidance for knowing when this will occur. By essentially adopting the idea of a fixed resource stock, these models ignore (or downplay) the underlying forces of technological change, which will influence exploration and extraction possibilities in the future, but also the importance of certain minerals for satisfying human needs. The peak advocates also tend to assume that oil and certain minerals are essential commodities in themselves, but the demand for these is a derived demand for a number of underlying human needs (food, shelter, travel etc.). To the extent that these needs can be satisfied with other more abundant resources, mineral depletion is not in itself a problem (even in the presence of a peak or decline in production).

Another important corollary of the peak theories is that the emerging economies of the world will not be given the opportunity to develop economically and socially using metals and minerals as key inputs. This route has been successfully used by the industrialized countries of today during the nineteenth and twentieth centuries. Recent developments in China, India and other developing countries show that this route is also being followed today, and that metals and minerals have not lost their importance in spite of technological progress bringing new materials and processes. This does not preclude that the emerging economies can develop more material more efficiently than has been the case in the past. As long as the peak theories are only discussed among academics, they will not cause much harm. But if the peak theorists manage to gain support from politicians and political action was taken based on these unproven theories, damage could be done. Today's emerging economies would perhaps be forced to try to find new, so far unknown, routes for their future economic development. They would not be able to use metals and minerals for their quick and efficient economic development. Exactly what political and social implications this will have is difficult to imagine but when it becomes clear that the peak theory was not sufficient to predict *when* metals are becoming seriously scarce, it is certain that the result could be serious. The mistrust between developed and emerging countries, between the haves and have-nots could deepen further and to rebuild it will take long.

In addition, the peak models do not provide us with much policy guidance either (Söderholm 2003). The reason is simply that the models do not explain thoroughly enough why we may witness a peak in production and, if any anticipated peak would indeed be due solely to geological constraints, this would inevitably be reflected in rising market prices. Thus, in the case of non-renewable resource extraction there is essentially no market failure arising from the pure fact that the resource is exhaustible, and peak modellers would therefore do more good to societal development if they instead paid attention to the goods and services that may not be properly internalized in the price mechanism (e.g. environmental pollution).

Our scepticism towards using peak approaches as a guide to the future should not be interpreted to suggest that research on the long-run availability of minerals is unimportant. On the contrary, the importance of metals and energy minerals for the global economy implies that such research is worthwhile pursuing. This research could address not only how resource scarcities could influence political conflicts and economic developments but also how conflicts may influence resource availability and shortages in the short and the long run. These types of shortages typically arise as a surprise (as a result of wars, embargos etc.), they can be quite costly and disruptive while they last, but so far they have typically persisted for only a few years.

Notes

1. Aleklett and Campbell (2003) even suggest that the 'transition [...] will include moves to a more central style of government as the famous open market is not designed to manage depletion of a critical resource' (p. 19).

2. This differs from the definition of reserves, which combines geological knowledge and the economic dimension (as in the so-called McKelvey diagram).
3. It can also be noted that the date of the global oil peak has been pushed further into the future when the year 2000 passed.
4. Of course, the ability to recycle – given existing technology – will vary across materials. For instance, the addition to lead in gasoline distributes lead in such a manner that its recovery is uneconomic. Moreover, some products contain multiple metals in ways that current technologies cannot reconstitute without impurity. Still, similar to the extraction of virgin resources, the above is not necessarily a sign of an impending peak in metal recycling.

Conclusion: Reconceptualizing the Dynamics of Conflict and Cooperation

Wojciech Ostrowski

From the mid-1980s until the beginning of the 2000s, the issue of oil, gas and minerals was largely absent from the global political agenda. The sharp decline in commodity prices during the 1990s, the collapse of the Soviet Union and the widespread triumph of neo-liberal policies meant that these resources were mainly treated within an economic framework where markets were dominant. In those occasions when these resources did feature in the political debates this was mostly framed as a development issue, such as with the resource-curse thesis, or was discussed in relation to civil wars that had erupted since the end of the Cold War. The rise of China and other emerging powers, the sharp increase in the commodity price in the mid-2000s and the increasing intervention of states in oil, gas and mineral markets put an end to this somewhat depoliticized understanding of resources. The outcome of this trend was the consolidation of two competing views concerning the future of global resources, framed by Dannreuther as between a realist as against a liberal model or narrative. This volume contributes to this debate by seeking to take a step back and to view the issue of conflict and collaboration from a multiplicity of different angles and perspectives rather than trying to construct a clear-cut discourse which would argue in favour of one narrative over another. In short, it sets out to problematize the issue at hand rather than to provide ammunition for either camp.

Issues of conflict and collaboration do not arise from the mere physical qualities of oil, gas and minerals but from the subsequent processes of extraction, production, transportation and sale of hydrocarbons and minerals on the local, regional and international markets. This process of mobilizing resources is enabled by a complex web of industrial, economic and political relations as well as various institutions that operate on both national and international levels. The relationship between different elements of this matrix is fluid and prone to change. For this reason the map of conflict and cooperation in relation to resources is complex and multidimensional. The matter is further complicated by the fact that different extractive industries share some similarities but also many dissimilarities with each other.

This book has sought to demonstrate that, nevertheless, these differences between global resources should not serve as a deterrent against a comparative study between different hydrocarbons and minerals. On the contrary, the introduction to this volume argued that a comparative study of oil, gas and minerals, set within a broader historical, political, economic and intellectual context, brings to the surface dynamics which can be lost in a more narrowly defined study. In addition, a wide-ranging comparative study also allows us to identify gaps in knowledge which researchers can address in the future.

It has been said many times before and restated by a number of contributors in this book: resources have fuelled the construction of the modern world but also created its limits. In turn, the modern political and economic systems, which these global resources enabled, have themselves shaped the environment in which conflict and cooperation over resources takes place. The key events in modern history such as European imperial expansion, the First World War, the Bolshevik Revolution, decolonization, the collapse of the USSR and now the rise of China, all constitute some of the key critical junctures in the history of resources (Dannreuther). Yet, as the preceding discussions in various chapters demonstrated, the history of resources is not driven solely by global 'external' events. The geographical location of these resources (Humphreys; Chindo and Bradshaw), the interplay between various private and state-owned extracting companies (Stevens; Luciani), complex infrastructure webs (Cragg), technology (Criqui; Bergmann; Ericsson and Söderholm), state-community relations and environmental concerns (Chindo and Bradshaw), legal frameworks (Dietsche), cartels, regional and international institutions (Dannreuther), state–society relationships (Ostrowski) and arguably most importantly the relative price of these resources (Stevens, Humphreys, Cragg) constitute a critical part of the space in which conflict and cooperation in relation to resources occurs. The intellectual debates in the fields of geopolitics, security studies and political economy (including the peak debates) frame the way in which we think about resources and as such academic discourses cannot be treated as value-free but rather as a constitutive part of the conflict/collaboration equation (Dannreuther; Ostrowski; Criqui; Bergmann; Ericsson and Söderholm).

Resources, actors and conflict and collaboration

One common theme in this book is how two key actors are particularly critical. These are states and companies and how their complex interrelations and interactions underline key conflicts and collaborations in relation to oil, gas and minerals. Both actors include a wide variation and are, importantly, continually evolving. Hence, in this volume when we talk about the state we do not just mean a resource-producing state but rather three families of states: (a) Western states; (b) postcolonial states; (c) post-revolutionary states (e.g. the Soviet Union and China). Similarly, by companies we did not solely mean Western transnational corporations (TNCs) but also National Oil Companies (NOCs), state-owned mining and gas companies (also known as national champions), independents and service companies. As Luciani highlights in his chapter, how states and companies have been interacting with underwent far-reaching changes over the twentieth and twenty-first centuries.

In the case of oil, the nexus between Western states and TNCs evolved into one between postcolonial states and TNCs and later on into an interplay between postcolonial state-NOCs, TNCs, independent companies and service companies (Stevens; Luciani). Similarly, in the case of mining a key relationship between Western states and TNCs eventually gave way to postcolonial state–TNCs interaction only to evolve into a complex relationship between postcolonial states and state-owned companies, TNCs and companies from China and other rising powers. As Humphreys notes, the key difference between oil and mineral industries is in the greater presence of TNCs throughout the history of the mining industry. The power of major oil companies failed to recover from the nationalization of the 1970s even during the neo-liberal era of the following two decades. In the case of gas, private companies generally took the lead and the sector developed with little involvement from the state. Yet, the state played a vital role in the construction of the pipeline system without which the gas industry would not have developed (Cragg).

The interaction between states and oil, mining and gas companies produced a number of conflicts on the international, regional and local levels. However, as has emerged in this book, some forms of conflicts are more specific to one resource rather than to others. Oil has been linked to conflicts that occur on the international level; gas is mostly associated with conflicts on the regional level and minerals are particularly linked to conflicts on the local level. The key triggers for these conflicts include, for example, oil price volatility, gas pipeline disputes and tensions between extractive companies and local communities. In general, such conflicts were subsequently followed by periods of cooperation between the different actors, attempting, often to mixed effect, to provide a durable long-term solution to these problems. For example, the issue of oil price volatility was first addressed by the cartel of the Western companies (Seven Sisters) and then subsequently by key oil-producing states (OPEC) but with decreasing effectiveness (Stevens; Luciani). The problems associated with pipeline disputes were minimized through the construction of additional pipelines that bypassed 'problematic' states (Cragg). In recent times, the issue of company–communities relations has been dealt with through initiatives such as Corporate Social Responsibility (CSR) (Chindo and Bradshaw). In short, conflicts over resources are generally balanced by acts of cooperation. In effect, as different contributors to this volume have demonstrated, conflict over oil and gas rarely endures over a prolonged period of time, and this is even less the case for gas which appears to be least conflict-prone of all the three resources discussed in this book.

Conflicts over oil and mineral resources first fully came into political focus in the 1970s in the aftermath of the oil shocks and the nationalization of a number of key extractive industries. Different and often disconnected developments in various parts of the world came to be seen as part of a larger assault against the North and its most profitable companies (Dannreuther; Ostrowski). A unifying feature of these conflicts was their geographical location. Almost all of them occurred in the postcolonial world. The oil and mineral industries which initially developed in the global North – together with the technology and the financial and legal frameworks (Dietsche) that serviced them – expanded to the global South. Political developments in the South in the postcolonial era greatly affected the oil and mineral industries. The fact that so many conflicts over oil and minerals have been connected to the postcolonial world goes some way to explain why the gas industry has been viewed as relatively conflict free. It is the only industry that took its initial steps in the global North and which developed firmly in Western Europe, and only later on in the USSR. By the same token today, the gas revolution is not taking place in the global South but in the US. In effect, the conflicts related to gas have been seen as less critical, apart from the Russian–Ukrainian dispute in the second half of 2000s, than those related to oil or minerals. Another critical factor responsible for gas being a less controversial resource has to do with its price; historically gas failed to produce rent close to the one generated by oil (Cragg).

The way forward: Approaches, regions and disciplinary boundaries

The events of the 1970s had a profound impact on the way in which we conceptualize and understand the role of oil, gas and minerals and resources. This period gave birth to various intellectual constructs such as energy security, the 'oil weapon' and the concept of resource nationalism. These concepts are regularly reactivated each time resources returns to the political arena, such as in the second half of the 2000s (Stevens; Humphreys; Dannreuther; Luciani). Yet these concepts in themselves are rarely critically assessed. For example, such ways of thinking drawn from the context of the 1970s contributed to the 'crisis' over rare earth minerals and which, as Humphreys and Ericsson and Söderholm note, was generally exaggerated and misunderstood. This example illustrates the fact that approaches drawn from the past can be potentially harmful and can obscure the complex reality of current developments which differ a great deal from the time of postcolonial struggles and the Cold War. However, as this volume shows, significant advances in our intellectual knowledge have been made. For example, the extensive literature on the issue of peak oil (and gas and minerals) demonstrates how vital assumptions regarding the geological depletion of resources and the far-reaching political, economic and social consequences of this process have been a topic of a serious and wide-ranging debate (Criqui; Bergmann; Ericsson and Söderholm). Furthermore, rentier state theory, which is still one of the key analytical tools within political science concerning resource-producing states, has been usefully and fruitfully applied to various parts of the world beyond the Middle East (Ostrowski).

However, the overwhelming focus on the ex-colonies of major Western states means that others areas of the world have been neglected. The Soviet Union, a major producer of oil, gas and minerals, is often portrayed as almost a marginal player or an entity which simply exited the history of resources from 1917 to 1991. While this is largely due to the fact that it is difficult, if not impossible, to conduct an in-depth study of the USSR extractive industries, it does not mean that this industry did not have a profound impact on the way in which a substantial amount of global resources were extracted, produced, transported and sold. Thus a new history of resources should look both at the ex-colonial world and at the Soviet Union, bestowing on them equal attention whenever possible. In addition, a better understanding of the role of the Soviet planners, the complex bargaining processes between Moscow and the peripheries and energy relations between the Soviet Union and Central and Eastern Europe would allow us to more fully comprehend the events of the 1990s and at the present time.

The history of the communist world also matters because the current rising powers were shaped by a communist rather than a Western experience. Indeed, the fact that state capitalism is a viable option for both the ex-revolutionary states of Russia and China means that the politics of resources is currently being forged by new and very different forces, which are best understood in their historical context. Essentially, therefore, we should reassess the impact of non-Western and non-colonial histories. For instance, the Bolshevik revolution is often mentioned in relation to the nationalization of the extractive industries in North Africa and the Middle East in the 1960s and 1970s. However, the Bolshevik revolution and the legacy of the total nationalization that followed have had an equally important effect on the emergence of the contemporary phenomenon of state capitalism. Indeed, taking into consideration both the past and present, it may well be that the Bolshevik revolution had a much more profound and long-term impact on the global politics of resources than the nearly mythical switch by the British navy from coal to oil.

Another important issue for future studies of resources and equally for the issue of conflict and collaboration is the question of disciplinary boundaries. Stevens and Luciani highlight in their chapters how in recent decades the relationship between the oil and gas industry and the global finance sector has become extremely close. As a result, financial intermediaries have become increasingly important in defining an energy company's strategy. Thus, the study of finance, which traditionally has often been ignored in studies of resources, should play a much larger role. By the same token, studies that deal with tax havens could be better integrated into the area of resource studies. Similarly, as against the relentless focus on the pathologies of the resource-producing states, more analysis should be accorded to how the flow of petrodollars destabilized the economies of Western and other countries, including resurrecting studies from the 1970s that examined the recycling of petrodollars and the links between extractive industries and the international arms trade.

Finally, this volume demonstrated that before arguing over the relative power of states and markets or adopting realist or liberal standpoints, we should first and foremost familiarize ourselves with the history of the resources in all their complexity. At the same time, this also requires a recognition, as this volume demonstrates, that history gives us no simple answers to how to understand our own transition from the existing energy supply system to a new one. As such, there are no timeless lessons to be taken from the transition from coal to oil for our own post-carbon predicament. The only fact that is certain is that the process will be long and complex and that this new reality will have an enormous impact not only on the economy but also politics. Timothy Mitchell (2011) noticed that the coal age fostered democratic politics by encouraging the emergence of organized trade labour movement. In the oil age that followed, new, more capital-intensive, geographically dispersed centres of energy production emerged, increasing the power of authoritarian governments. How political systems on the national and international levels will be arranged in the post-carbon age is an open-ended question.

Bibliography

AccountAbility and BSR (Business for Social Responsibility) (2004) *Business and Economic Development: Mining Sector Report*, London: AccountAbility and Business for Social Responsibility.

Acemoglu, D. (2003) 'Why not a Political Coase Theorem? Social Conflict, Commitment and Politics', *Journal of Comparative Economics*, 31(5): 620–52.

Acemoglu, D. and Johnson, S. (2005) 'Unbundling Institutions', *Journal of Political Economy*, 113(5): 949–95.

Acemoglu, D. and Robinson, J. (2012) *Why Nations Fail: The Origins of Power, Prosperity and Poverty*, New York: Crown.

Achebe, C. H., Nneke, U. C. and Anisiji, O. E. (2012) 'Analysis of Oil Pipeline Failures in the Oil and Gas Industries in the Niger Delta Area of Nigeria', *Proceedings of the International Multi Conference of Engineers and Computer Scientist Vol. II*, Hong Kong.

Adegoke, O. S. (1974) 'Preliminary Proposal for the Exploration and Utilisation of Tar Sands of Western State of Nigeria', *Geology Consultancy Unit*, University of Ife, unpublished.

Adegoke, O. S. and Ibe, E. C. (1982) 'The Tar Sand and Heavy Crude Resources of Nigeria', *Proceedings of 2nd International Conference on Heavy Crude and Tar Sands*, Caracas, Venezuela.

Adelman, M. A. (1972) *The World Petroleum Market*, Baltimore, MD: Johns Hopkins University Press.

Adelman, M. A. (1990) 'Mineral Depletion, with Special Reference to Petroleum', *The Review of Economics and Statistics*, 7(1): 1–10.

Adelman, M. A. (1993) *Economics of Petroleum Supply*, Cambridge, MA: MIT Press.

Adelman, M. A. (2004) 'The Real Oil Problem', *Regulation*, 27(1): 16–21.

AEA (2012) 'Support to the Identification of Potential Risks for the Environment and Human Health Arising from Hydrocarbons Operations Involving Hydraulic Fracturing in Europe', *Report for the European Commission*. Available at http://ec.europa.eu/environment/integration/energy/pdf/fracking%20study.pdf [Accessed 25 October 2012].

Aguilera, R. F., Eggert, R. G., Lagos, G. C. and Tilton, J. E. (2009) 'Depletion and the Future Availability of Petroleum Resources', *The Energy Journal*, 30(1): 141–74.

Aissaoui, A. (2001) *Algeria: The Political Economy of Oil and Gas*, Oxford: Oxford University Press, Oxford Institute for Energy Studies.

Aissaoui, A. (2011) 'The Changing Role and Relationship of IOCs, NOCs, and OFSCs and how it is Perceived by our Readers?' *APICORP Economic Commentary*, 6(8): 1–3.

Aleklett, K. and Campbell, C. J. (2003) 'The Peak and Decline of World Oil and Gas Production', *Minerals and Energy*, 18(1): 5–20.

Aleklett, K, Hööka, M., Jakobssona, K., Lardellib, M., Snowdenc, S. and Söderbergha, B. (2010) 'The Peak of the Oil Age – Analysing the World Oil Production Reference Scenario in World Energy Outlook 2008', *Energy Policy*, 38(3): 1398–414.

Alexander, M. (1990) 'Reclamation after Tin Mining on the Jos Plateau Nigeria', *The Geographical Journal*, 156(1): 44–50.

Al-Moneef, M. A. (1998) 'International Downstream Integration of National Oil Companies', in P. Stevens (ed.), *Strategic Positioning in the Oil Industry: Trends and Options*, London: I.B. Tauris.

Alston, L. J., Eggertsson, T. and North, D. C. (eds) (1996) *Empirical Studies in Institutional Change*, Cambridge: Cambridge University Press.

Amnesty International (2009) *Nigeria: Petroleum, Pollution and Poverty in the Niger Delta*, London: Amnesty International Publications.

Anderson, I. H. (1981) *ARAMCO: The United States and Saudi Arabia*, Princeton, NJ: Princeton University Press.

Anderson, L. (1987) 'The State in the Middle East and North Africa', *Comparative Politics*, 20(1): 1–18.

Andrew, H. and Jones, B. (2010) 'How do Rising Powers Rise?' *Survival*, 52(6): 63–88.

Andrews-Speed, P. and Dannreuther, R. (2011) *China, Oil and Global Politics*, London: Routledge.

Antill, N. and Robert, A. (2002) *Oil Company Crisis – Managing Structure, Profitability and Growth*, Oxford: Oxford Institute for Energy Studies.

API (American Petroleum Institute) (2009) *HF1 – Hydraulic Fracturing Operations – Well Construction and Integrity Guidelines*, 1st edn, October. Available at http://www.api.org/policy-and-issues/hf.aspx [Accessed 25 October 2012].

API (American Petroleum Institute) (2010a) *HF2 – Water Management Associated with Hydraulic Fracturing*, 1st edn, June. Available at http://www.api.org/policy-and-issues/hf.aspx [Accessed 25 October 2012].

API (American Petroleum Institute) (2010b) *Std 65 Part 2 – Isolating Potential Flow Zones During Well Construction*, 2nd edn, December. Available at http://www.api.org/policy-and-issues/hf.aspx [Accessed 25 October 2012].

API (American Petroleum Institute) (2011) *American Petroleum Institute. HF3 – Practices for Mitigating Surface Impacts Associated with Hydraulic Fracturing*, 1st edn, February. Available at http://www.api.org/policy-and-issues/hf.aspx [Accessed 25 October 2012].

Appiah-Adu, K. (1999) 'The Impact of Economic Reform on Business Performance: A Study of Foreign and Domestic Firms in Ghana', *International Business Review*, 8: 463–86.

Arnott, R. (2005) 'The Private Oil Companies: From Consolidation to Growth', *Oxford Energy Forum*, 60: 10–11.

Aroh, K. N., Ubong, I. U., Eze, C. L., Harry, I. M., Umo-Otong, J. C. and Gobo, A. E. (2010) 'Oil Spill Incidents and Pipeline Vandalization in Nigeria: Impact on Public Health and Negation to Attainment of Millennium Development Goal: The Ishiagu Example', *Disaster Prevention and Management*, 19(1): 70–87.

Attanasi, E. and Meyer, R. (2007) 'Natural Bitumen and Heavy Oil', *Survey of World Energy Resources*, The World Energy Council. Available at http://bit.ly/axjUlu [Accessed 19 July 2009].

Auty, R. (1990) *Resource-Based Industrialisation: Sowing the Oil in Eight Developing Countries*, New York: Oxford University Press.

Auty, R. (1993) *Sustaining Development in Mineral Economies: The Resource Curse Thesis*, London: Routledge.

Auty, R. (1995) *Patterns of Development: Resources, Policy and Economic Growth*, London: Edward Arnold.

Auty, R. (eds) (2001) *Resource Abundance and Economic Development*, Oxford: Oxford University Press.

Ayoade, E. (2007) 'Bitumen Resources of Nigeria', *Status Report*, Senate Committee on Solid Minerals, Abuja: Nigeria.

Ayubi, N. (2001) *Over-Stating the Arab State: Politics and Society in Middle East*, London: I.B. Tauris Publishers.

Azapagic, A. (2004) 'Developing a Framework for Sustainable Development Indicators for the Mining and Minerals Industry', *Journal of Cleaner Production*, 12: 639–62.

Babusiaux, D. and Bauquis, P.-R. (2007) 'Que penser de la raréfaction des ressources pétrolières et de l'évolution des prix du brut?' *Les cahiers de l'économie*, 66: 1–39.

Baker Institute (2007) 'The Changing Role of National Oil Companies in International Energy Markets', *Baker Institute Policy Report 35*, The James A. Baker III Institute for Public Policy of Rice University.

Baker, S. (2011) *Coal Mining: The Rise and Fall of a Once-Great Industry*. Available at http://www.screenonline.org.uk/history/id/1198219/index.html [Accessed 12 October 2011].

Bala-Gbogbo (2011) 'Nigeria's Oil Revenue Rose 46% to $59 Billion in 2010 on Improved Security', *Bloomberg*. Available at http://www.bloomberg.com/news/2011-04-14/nigeria-s-oil-revenue-rose-46-to-59-billion-in-2010-on-improved-security.html [Accessed 20 November 2011].

Baldwin, D. A. (ed.) (1993) *Neorealism and Neoliberalism: The Contemporary Debate*, New York: Columbia University Press.

Bamat, T. (1977) 'Relative State Autonomy and Capitalism in Brazil and Peru', *Critical Sociology*, 7(2): 74–84.

Bamberg, J. H. (1994) *The History of the British Petroleum Company: Volume 2, The Anglo-Iranian Years, 1928–1954*, Cambridge: Cambridge University Press.

Bardi, U. and Pagani, M. (2007) *Peak Minerals, The Oil Drum: Europe*. Available at http://www.theoildrum.com/node/3086 [Accessed 2 October 2012].

Barnes, J. and Chen, M. E. (2007) 'NOCs and US Foreign Policy', *Policy Report*, The James A. Baker Institute for Public Policy, Rice University, March.

Barry, A. (2006) 'Technological Zones', *European Journal of Social Theory*, 9(2): 239–53.

Barty-King, H. (1985) *New Flame: How Gas Changed the Commercial, Domestic and Industrial Life in Britain from 1783 to 1984*, Tavistock: Graphmitre.

Bastida, A. E. (2001) 'A Review of the Concept of Security of Mineral Tenure: Issues and Challenges', *The CEPMLP (Centre for Petroleum, Minerals Law and Policy) Internet Journal*. Available at http://bit.ly/e8WVho [Accessed 12 April 2009].

Bastida, A. E. (2004) 'Mining Law: New Directions?' in A. E. Bastida, T. Waelde and J. W. Fernandez (eds), *International and Comparative Mineral Law & Policy: Trends and Prospects*, The Hague: Kluwer Law International.

Bastida, A. E. (2006) 'Sustainable Investment in the Minerals Sector: Re-Examining the Paradigm', *International Environmental Agreements*, 6(4): 401–6.

Bastida, A. E. (2008) 'Mining Law in the Context of Development', in P. Andrews-Speed (ed.), *International Competition for Resources: The Role of Law, State and Markets*. Dundee: Dundee University Press.

Bates, R. H. (1989) *Beyond the Miracle of the Market: The Political Economy of Agrarian Development in Kenya*, Cambridge: Cambridge University Press.

Bates, R. H. (2001) *Prosperity and Violence: The Political Economy of Development*, New York: W. W. Norton & Company.

Bates, R. H. (2008) *When Things Fell Apart: State Failure in Late 20th Century Africa*, Cambridge: Cambridge University Press.

Bayulgen, O. (2010) *Foreign Investment and Political Regimes: The Oil Sector in Azerbaijan, Russia, and Norway*, Cambridge: Cambridge University Press.

BBC (2010) *UK Court Action Over Congolese 'Illegal Minerals'*, 26(July). Available at http://www.bbc.co.uk/news/uk-10767692 [Accessed May 2012].

Beblawi, H. (1987) 'The Rentier State in the Arab World', in H. Beblawi and G. Luciani (eds), *The Rentier State: Volume II*, London: Croom Helm.

Benner, T. and Soares de Oliveira, R. with Kalinke, F. (2010) 'The Good/Bad Nexus in Global Energy Governance', in Andreas Goldthau and Jan Martin Witte (eds), *Global Energy Governance: The New Rules of the Game*, Washington, DC: Brookings Institution Press.

Bentham, R. W. (1988) 'Legal Status of State Petroleum Companies', in N. Beredjick and T. Walde (eds), *Petroleum Investment Policies in Developing Countries*, London: Graham & Trotman.

Bentham, R. W. and Smith, W. G. R. (1986) *State Petroleum Corporations: Corporate Forms, Powers and Control*, University of Dundee: Centre for Petroleum and Mineral Law and Policy.

Bentley, R. W. (2002) 'Global Oil and Gas Depletion: An Overview', *Energy Policy*, 30: 189–205.

Berkowitz, D., Pistor, K. and Richard, J.-F. (2003) 'The Transplant Effect', *American Journal of Comparative Law*, 51(11): 163–204.

Bindemann, K. (1999) *Production-Sharing Agreements: An Economic Analysis*, Oxford: Oxford Institute for Energy Studies.

Blair, J. M. (1976) *Control of Oil*, New York: Macmillan.

Blank, S. (1995) 'Energy, Economics and Security in Central Asia: Russia and its Rivals', *Central Asian Survey*, 14(3): 373–406.

Bloomberg (2012) *Fracking Needs Rules, Not Flawed Studies*, 21(October). Available at http://www.bloomberg.com/news/2012-10-21/fracking-needs-rules-not-flawed-studies.html [Accessed 25 October 2012].

Bouè, J. C. (1993) *Venezuela: The Political Economy of Oil*, Oxford: Oxford University Press, Oxford Institute for Energy Studies.

Bouè, J. C. (2004) *The Internationalisation Programme of PDVSA*, Oxford: Oxford Institute for Energy Studies.

Bowler, P. A. (1948) 'The Mining Industry: The Nigerian Tin Mining-Industry', in M. Perham (ed.), *Mining, Commerce, and finance in Nigeria: Being the Second Part of a Study of the Economics of a Tropical Dependency*, London: Faber and Faber.

Boyd, L. (1994) 'The Economics of the Coal Company Town: Institutional Relationships, Monophony, and Distributional Conflicts in American Coal Towns', *The Journal of Economic History*, 54(20): 426–7.

BP (2010) *BP Statistical Review*, London: British Petroleum.

BP (2011) *Energy Outlook 2030*, London: British Petroleum.

BP (2012) *Statistical Review of World Energy*, London: British Petroleum.

Bradley, R. L. (1996) *Oil, Gas and Government: The US Experience, Volumes I and II*, Maryland: Rowman & Littlefield.

Brandt, W. (1980) *North-South: A Programme for Survival (Report of the Independent Commission on International Development Issues)*, London: Pan Books.

Bratton, M. and van de Walle, N. (1994) 'Neopatrimonial Regimes and Political Transitions in Africa', *World Politics*, 46(4): 453–89.

Bräutigam, D., Fjeldstad, O. -H. and Moore, M. (eds) (2008) *Taxation and State Building in Developing Countries: Capacity and Consent*, Cambridge: Cambridge University Press.

Bray, J. (2003) 'Attracting Reputable Companies to Risky Environments: Petroleum and Mining Companies', in I. Bannon and P. Collier (eds) *Natural Resources and Violent Conflict: Options and Actions*, Washington, DC: The World Bank.

Bremmer, I. (2009) 'State Capitalism comes of Age: The End of the Free Market?' *Foreign Affairs*, 88(3): 40–55.

Bremmer, I. (2010) *The End of the Free Market: Who Wins the War Between States and Corporations?* London: Portfolio.

Bridge, G. (2000) 'The Social Regulation of Resources Access and Environmental Impact: Production, Nature and Contradiction in the US Copper Industry', *Geoforum*, 31(2): 237–56.

Bridge, G. (2004a) 'Contested Terrain: Mining and the Environment', *Annual Review of Environmental Resources*, 29: 205–59.

Bridge, G. (2004b) 'Mapping the Bonanza: Geographies of Mining Investment in an Era of Neoliberal Reform', *The Professional Geographer*, 56(3): 406–21.

Bridge, G. (2008a) 'Global Production Networks and the Extractive Sector: Governing Resource-Based Development', *Journal of Economic Geography*, 8(3): 389–419.

Bridge, G. (2008b) 'Economic Geography: Natural Resources', in R. Kitchin and N. Thrift (eds) *International Encyclopaedia of Human Geography*, London: Elsevier.

Brumberg, D. and Ahram, A. I. (2007) 'The National Iranian Oil Company in Iranian Politics', *Policy Report*, The James A. Baker Institute for Public Policy, Rice University.

Brynen, R. (1992) 'Economic Crisis and Post-Rentier Democratization in the Arab World: The Case of Jordan', *Canadian Journal of Political Science*, 25(1): 69–97.

Bureau of Infrastructure, Transport and Regional Economics (BITRE) (2009), 'Transport Energy Futures: Long-Term Oil Supply Trends and Projections', *Report 117*, Canberra, ACT.

Buzan, B., Waever, O. and de Wilde J. (1998) *Security: A New Framework for Analysis*, London: Lynne Rienner.

Cameron, P. D. (2010) *International Energy Investment Law: The Pursuit of Stability*, Oxford: Oxford University Press.

Campbell, B. (ed.) (2004) 'Regulating Mining in Africa: For Whose Benefits?' *Discussion Paper 26*, Nordiska Afrikainstitutet, Uppsala.

Campbell, B. (2006) 'Good Governance, Security and Mining in Africa', *Minerals and Energy – Raw Materials Report*, 21(1): 31–44.

Campbell, B. (2010) 'Revisiting the Reform Process of African Mining Regimes', *Canadian Journal of Development Studies*, 30(1–2): 197–217.

Campbell, C. and Laherrère, J. (1998) 'The End of Cheap Oil', *Scientific American*, March. Available at http://dieoff.org/page140.pdf [Accessed 2 October 2012].

Canadian Centre for Energy Information (2009) *Oil Sands and Heavy Oil Overview*, CCEI. Available at http://bit.ly/elbVBQ [Accessed 12 July 2009].

Carl, T. M. and Smith, M. B. (2010) 'Hydraulic Fracturing – History of an Enduring Technology', *Journal of Petroleum Technology*, December: 26–41. Available at http://www.spe.org/jpt/print/archives/2010/12/10Hydraulic.pdf [Accessed 25 October 2012].

Carnegy, H. (2012) 'Hollande Rejects Shale Gas Fracking', *Financial Times*, September 14, 2012.

Carr, E. H. (1946) *The Twenty Year's Crisis: An Introduction to the Study of International Relations*, London: Macmillan.

Castaneda, C. J. and Clarance, S. M. (1996) *Gas Pipelines and the Emergence of America's Regulatory State: A History of Panhandle Eastern Corporation, 1928–1993*, New York: Cambridge University Press.

Cavalcanti, C. (2000) 'Cleso Furtado and the Persistence of Underdevelopment', *Paper Read at the University of Oxford Centre for Brazilian Studies' Conference and Seminar Programme*, Michaelmas Term, 20th November 2000.

Cawthra, M. (2011) 'Collusion and Criminalisation: Fuel Conflict in the Niger Delta', in V. Gounden (ed.), *Conflict Trends 2011/12*, Umhlanga Rocks: African Centre for the Constructive Resolution of Disputes.

CERA (2009) 'The Future of Global Oil Supply: Understanding the Building Blocks', *CERA Special Report*, Cambridge, MA.

Chambers, A. (1999) *Natural Gas and Electric Power in Non-Technical Language*, Oklahoma: PennWell.

Chandra, V. (2006) *Fundamentals of Natural Gas: An International Perspective*, Oklahoma: PennWell.

Charpentier, A. D., Bergeson, J. A. and MacLean, H. L. (2009) 'Understanding the Canadian Oil Sand Industry's Greenhouse Gas Emissions', *Environmental Research Letters*, 4(014005): 1–11.

Chaudhry, K. A. (1989) 'The Price of Wealth: Business and State in Labour Remittance and Oil Economies', *International Organisation*, 43(1): 101–45.

Chen, M. E. (2007) 'National Oil Companies and Corporate Citizenship: A Survey of Transnational Policy and Practice', *Policy Report*, The James A. Baker Institute for Public Policy, Rice University.

Cheney, H., Lovel, R. and Solomon, F. (2002) 'People, Power, Participation: A Study of Mining – Community Relationships', *Minerals Mining and Sustainable Development Project*, International Association of Environment and Development.

Chevalier, J. -M. (1973) *The New Oil Stakes*, Wappingers Falls, NY: Beekman Books.

Chirot, D. and Hall, T. (1982) 'World-System Theory', *Annual Review of Sociology*, 8: 81–106.

Clark, J. (1997) 'Petro-Politics in Congo', *Journal of Democracy*, 8(3): 62–76.

Clark, J. and Clark, A. (1996) 'The New Reality of Mineral Development: Social and Cultural Issues in Asia and Pacific Nations', *Resources Policy*, 25(3): 189–96.

Clay, K. and Wright, G. (2003) 'Order without Law? Property Rights during the Californian Gold Rush', *Working Paper No. 265*, Stanford Law School, Stanford.

Coase, R. (1960) 'The Problem of Social Cost', *Journal of Law and Economics*, 3: 1–44.

Collier, P. (2000) 'Doing Well Out of War: An Economic Perspective', in M. Berdal and D. Malone (eds), *Greed and Grievance: Economic Agendas in Civil Wars*, Boulder, CO: Lynne Rienner.

Collier, P. (2008) *The Bottom Billion*, Oxford: Oxford University Press.

Collier, P. and Hoeffler, A. (1998) 'On the Economic Causes of Civil War', *Oxford Economic Paper*, 50(4): 563–73.

Collier, P. and Hoeffler, A. (2002) 'Greed and Grievance in Civil Wars', *Working Paper*, Oxford: Centre for the Study of African Economies.

Collier, P. and Hoeffler, A. (2004) 'Greed and Grievance in Civil War', *Oxford Economic Papers*, 56: 563–95.

Cotula, L. (2008) 'Reconciling Regulatory Stability and Evolution of Environmental Standards in Investment Contracts: Towards a Rethinking of Stabilization Clauses', *World Energy Law and Business*, 1(2): 158–79.

Council on Economics and National Security (1981) *Strategic Minerals: A Resource Crisis*, Washington, DC: US Council on Economics and National Security.

Cox, R. W. (1981) 'Social Forces, States and World Orders: Beyond International Relations Theory', *Millennium*, 10(2): 126–55.

Cramton, P. (2007) 'How Best to Auction Oil Rights', in M. Humphreys, J. D. Sachs and J. E. Stiglitz (eds), *Escaping the Resource Curse*, New York: Columbia University Press.

Criqui, P. and Mima, S. (2012) 'European Climate – Energy Security Nexus: A Model Based Scenario Analysis', *Energy Policy*, 41(1): 827–42.

Crowson, P. C. F. (2008) *Mining Unearthed*, London: Aspermont UK.

Crystal, J. (1994) 'Authoritarianism and its Adversaries in the Arab World', *World Politics*, 46(2): 262–89.

Currie, J., Greely, D., Nathan, A., Serio, G., Dart, S., Zhang, R. and Khan, A. (2008) *The Revenge of the Old 'Political' Economy*, Goldman Sachs. Available at http://www.fullermoney.com/content/2008-05-07/CurrieRevengeoftheOldPoliticalEconomy.pdf [Accessed May 2012].

Daintith, T. (2010a) 'The Rule of Capture: The Least Worst Property Rule for Oil and Gas', in A. McHarg, B. Barton, A. Bradbrook and L. Godden (eds), *Property and the Law in Energy and Natural Resources*, Oxford: Oxford University Press.

Daintith, T. (2010b) *Finders Keepers? How the Law of Capture Shaped the World Oil Industry*, London: Resources For the Future Press/Earthscan.

Dalby, S., Routledge, P. and Toal, G. (eds) (2006) *The Geopolitics Reader*, London: Routledge.

Daniels, P. (1995) 'Evaluating State Participation in Mineral Projects: Equity, Infrastructure and Taxation', in J. Otto (ed.), *The Taxation of Mineral Enterprises*, London: Kluwer Law International.

Dargay, J., Gately, D. and Sommer, M. (2007) 'Vehicle Ownership and Income Growth, Worldwide: 1960-2030', *Energy Journal*, 28(4): 163–90.

Darley, J. (2009) *High Noon for Natural Gas: The New Energy Crisis*, Vermont: Chelsea Green Publishing Company.

Darmstadter, J., Teilelbaum, P. D. and Polach, J. G. (1971) *Energy in the World Economy*, Baltimore, MD: The Johns Hopkins Press for Resources for the Future Inc.

De Almeida, P. and Silva, P. D. (2009) 'The Peak of Oil Production – Timings and Market Recognition', *Energy Policy*, 37: 1267–76.

De Castro, C., Miguel, L. J. and Mediavilla, M. (2009) 'The Role of Non Conventional Oil in the Attenuation of Peak Oil', *Energy Policy*, 37(5): 1825–33.

De Oliveira, A. (2012) 'Brazil's Petrobras: Strategy and Performance', in D. G. Victor, David R. Hults and Mark Thurber (eds), *Oil and Governance: State-Owned Enterprises and the World Energy Economy*, Cambridge: Cambridge University Press.

De Soto, H. (2000) *The Mystery of Capital: Why Capitalism Triumphs in the West and Fails Everywhere Else*, London: Black Swan Books.

De Soysa, I. (2002) 'Paradise is a Bazaar? Greed, Creed, and Governance in Civil War, 1989–99', *Journal of Peace Research*, 39(4): 395–416.

Demsetz, H. (1967) 'Toward a Theory of Property Rights', *American Economic Review*, 57(3): 347–59.

Department of Energy (Canada) (1972) *An Energy Policy for Canada: Phase 1, Volume 1*, Ottawa: The Department Of Energy, Mines and Natural Resources.

Department of Energy and Resource Management (2011) *Issue Identification and Community Consultation*, Australia: Queensland Government. Available at http://www.derm.qld.gov.au/register/p00439aa.pdf [Accessed 29 May 2011].

Dogaru, D., Zobrist, J., Balteanu, D., Popescu, C., Sima, M. Amini, M. and Yang, H. (2009) 'Community Perception of Water Quality in a Mining-Affected Area: A Case Study for the Certej Catchment in the Apuseni Mountains in Romania', *Environmental Management*, 43(6): 1131–45.

Downs, E. (2008) 'Business Interest Groups in Chinese Politics: The Case of the Oil Companies', in Cheng Li (ed.), *China's Changing Political Landscape: Prospects for Democracy*, Washington, DC: Brookings Institution Press.

Drimmer, J. (2010) 'Human Rights and the Extractive Industries: Litigation and Compliance Trends', *Journal of World Energy Law and Business*, 3(2): 121–39.

Dumett, R. E. (1988) 'Sources for Mining Company History in Africa: The History and Records of the Ashanti Goldfields Corporation (Ghana), Ltd.', *The Business History Review*, 62(3): 502–15.

Dung-Gwom, J. Y. (2007) 'Post Mining Operations and the Environment', *38th Annual Conference of the Nigerian Institute of Town Planners, Solid Minerals Exploitation in Nigeria: Its Challenges to Land Use Planning*, Asaba, Nigeria.

Dunne, T., Kurki, M. and Smith, S. (eds) (2010) *International Relations Theories: Discipline and Diversity*, Oxford: Oxford University Press.

Dupuy, A. and Truchil, B. (1979) 'Problems in the Theory of State Capitalism', *Theory and Society*, 8(1): 1–38.

Earthworks and Oxfam (2004) *Dirty Metals: Mining, Communities and the Environment*, Washington, DC: Earthworks and Oxfam America.

Ebel, R. (eds) (2003) 'Caspian Oil Windfalls: Who will Benefit?' *Caspian Revenue Watch*, New York: Central Eurasia Project.

Eckes, A. (2011) *The Contemporary Global Economy: A History since 1980*, Oxford: Wiley-Blackwell.

Ecologist (1972) 'A Blueprint for Survival', 2(1).

Economist (2006) 'Never too Late to Scramble', 28(Oct.).

Economist (2007) 'What's Mined in Yours', 6(Oct.).

Economist (2011) 'Digging for Victory', 24(Sep.).

Economist (2012) 'The Rise of State Capitalism', *Economist*, 21(Jan.).

ECSSR (Emirates Center for Strategic Studies and Research) (eds) (2005) *Gulf Oil in the Aftermath of the Iraq war: Strategies and Policies*, Abu Dhabi: ECSSR.

Eggertsson, T. (1990) *Economic Behaviour and Institutions: Cambridge Surveys of Economic Literature*, Cambridge: Cambridge University Press.

Eggertsson, T. (2005) *Imperfect Institutions: Possibilities and Limits of Reform*, Ann Arbor, MI: University of Michigan Press.

EIA (Energy Information Administration) (2011a) *International Energy Outlook 2011*, Washington, DC.

EIA (Energy Information Administration) (2011b) 'Lower 48 States Shale Plays', 9(May). Available at http://www.eia.gov/oil_gas/rpd/shale_gas.pdf [Accessed 25 October 2012].

EIA (Energy Information Administration) (2011c) *World Shale Gas Resources: An Initial Assessment of 14 Regions Outside the United States*, Washington, DC. Available at http://www.eia.gov/analysis/studies/worldshalegas/ [Accessed 25 October 2012].

EIA (Energy Information Administration) (2012a) 'Glossary'. Available at http://www.eia.gov/tools/glossary/index.cfm?id=u [Accessed 25 October 2012].

EIA (Energy Information Administration) (2012b) *Annual Energy Outlook 2012*, Washington, DC. Available at http://www.eia.gov/forecasts/aeo/pdf/0383(2012).pdf [Accessed 25 October 2012].

EIA (Energy Information Administration) (2012c) 'Table 14 Unproved Technically Recoverable Resource Assumptions by Basin', *Annual Energy Outlook 2012*, Washington, DC. Available at http://www.eia.gov/forecasts/aeo/table_14.cfm [Accessed 25 October 2012].

EIA (Energy Information Administration) (2012d) 'Natural Gas Consumption by End Use'. Available at http://www.eia.gov/dnav/ng/ng_cons_sum_dcu_nus_a.htm [Accessed 25 October 2012].

EIA (Energy Information Administration) (2012e) 'Today in Energy: Natural Gas Prices near 10-Year Low Amid Mild Weather, Higher Supplies in Winter, 2011–12', 19(April). Available at http://www.eia.gov/todayinenergy/detail.cfm?id=5910# [Accessed 25 October 2012].

EIA (Energy Information Administration) (2012f) 'Today in Energy: US Natural Gas Net Imports at Lowest Levels since 1992', 15(March 15). Available at http://www.eia.gov/todayinenergy/detail.cfm?id=5410 [Accessed 25 October 2012].

EIA (Energy Information Administration) (2012g) 'Today in Energy: US Energy-Related CO_2 Emissions in Early 2012 Lowest since 1992', 1(August). Available at http://www.eia.gov/todayinenergy/detail.cfm?id=7350 [Accessed 25 October 2012].

EIA (Energy Information Administration) (2012h) *Country Analysis Briefs: Nigeria*, Washington, DC.

Eifert, B., Gelb, A. and Tallroth, N. B. (2003) 'The Political Economy of Fiscal Policy and Economic Management in Oil-Exporting Countries', in J. M. Davis, R. Ossowski and A. Fedelino (eds), *Fiscal Policy Formulation and Implementation in Oil-Producing Countries*, Washington, DC: International Monetary Fund.

EirGrid (2010) *Annual Renewable Report 2010*. Available at http://www.eirgrid.com/media/Annual%20Renewable%20Report%202010.pdf [Accessed 15 October 2012].

Eller, S. L., Hartley, P. and Medlock, K. B. III (2007) 'Empirical Evidence on the Operational Efficiency of National Oil Companies', *Policy Report*, The James A. Baker Institute for Public Policy, Rice University.

Energy Institute (2012) 'Fact-Based Regulation for Environmental Protection in Shale Gas Development – Summary of Findings', University of Texas, February.

Entelis, J. P. (2012) 'Sonatrach: The Political Economy of an Algerian State Institution', in D. G. Victor, D. R. Hults and M. C. Thurber (eds), *Oil and Governance: State-Owned Enterprises and the World Energy Supply*, Cambridge: Cambridge University Press.

EPA (Environmental Protection Agency) (2011) 'DRAFT – Investigation of Ground Water Contamination near Pavillion, Wyoming', *National Risk Management Research Laboratory*, EPA 600/R-00/000, December. Available at http://www.epa.gov/region8/superfund/wy/pavillion/EPA_ReportOnPavillion_Dec-8-2011.pdf [Accessed 25 October 2012].

EPA (Environmental Protection Agency) (2012a) 'Natural Gas Extraction – Hydraulic Fracturing'. Available at http://www.epa.gov/hydraulicfracturing/ [Accessed 25 October 2012].

EPA (Environmental Protection Agency) (2012b) 'The Safe Drinking Water Act's (SDWA) Underground Injection Control Program'. Available at http://water.epa.gov/type/groundwater/uic/index.cfm [Accessed 25 October 2012].

EPA (Environmental Protection Agency) (2012c) 'Summary of the Clean Water Act'. Available at http://www.epa.gov/lawsregs/laws/cwa.html [Accessed 25 October 2012].

EPA (Environmental Protection Agency) (2012d) 'Oil and Natural Gas Air Pollution Standards'. Available at http://www.epa.gov/airquality/oilandgas/actions.html [Accessed 25 October 2012].

EPA (Environmental Protection Agency) (2012e) 'Proposed Amendments to Air Regulations for the Oil and Natural Gas Industry-Fact Sheet'. Available at http://www.epa.gov/airquality/oilandgas/pdfs/20110728factsheet.pdf [Accessed 25 October 2012].

ERCB (Energy Resources Conservation Board) (2009) *ST98-2009: Alberta's Energy Reserves 2008 and Supply/Demand Outlook 2009–2018*, Alberta: Energy Resources Conservation Board.

Ericsson, M. (2008) 'Seabed Deposits Generate a New Wave of Interest in Offshore Mineral Recovery', *Engineering and Mining Journal*, September: 120–5.

Esteves, A. (2008) 'Mining and Social Development: Refocusing Community Investment and using Multi-Criteria Decision Analysis', *Resources Policy*, 33: 39–47.

European Commission (2007) *DG Competition Report on Energy Sector Inquiry*, (SEC(2006)1724, January 10, 2007).

European Commission (2010) *Critical Raw Materials for the EU*, Brussels: Report of the Ad-hoc Working Group on Defining Critical Raw Materials.

European Commission (2011a) *A Roadmap For Moving to a Competitive Low Carbon Economy in 2050*, Brussels, 8.3.2011 COM(2011) 112.

European Commission (2011b) *Energy Roadmap 2050*, COM(2011) 885.

Evans, L. (2006) 'Science and Technology for End-Uses after Mining', *International Journal of Mining, Reclamation and Environment*, 20(2): 83–4.

Evans, P. B. (2007) 'Extending the "Institutional" Turn: Property, Politics, and Development Trajectories', in H.-J. Chang (ed.), *Institutional Change and Economic Development*, London: Anthem Press and UNU Press.

Farsoun, K. (1975) 'State Capitalism in Algeria', *Middle East Research and Information Project, No. 35.*

Fattouh, B. (2011) *An Anatomy of the Crude Oil Pricing System*, Oxford: Oxford Institute for Energy Studies.

Fattouh, B. and van den Linde, C. (2011) *The International Energy Forum: Twenty Years of Producers-Consumers Dialogue in a Changing World*, Riyadh: IEF.

Fearon, J. and David, L. (2003) 'Ethnicity, Insurgency, and Civil War', *American Political Science Review*, 97(1): 75–90.

Ferguson, J. (2008) 'Seeing like an Oil Company: Space, Security, and Global Capital in Neoliberal Africa', *American Anthropologist*, 107(3): 377–82.

Fernandez, R. and Ocampo, J. (1975) 'The Andean Pact and State Capitalism in Columbia', *Latin American Perspectives*, 2(3): 19–35.

Firmin-Sellers, K. (1996) *The Transformation of Property Rights in the Gold Coast: An Empirical Study Applying Rational Choice Theory*, Cambridge: Cambridge University Press.

Fonseca, H. (ed.) (2004) *Mining: Social and Environmental Impact*, Uruguay: World Rainforest Movement.

Forbes Alex (2010) 'The 40 Million Barrel Question', *European Energy Review*, 7(May).

Fouquet, R. and Pearson, P. J. G. (1998) 'A Thousand Years of Energy Use in the United Kingdom', *Energy Journal*, 19(4): 1–41.

Franke, A., Gawrich, A. and Alakbarov, G. (2009) 'Kazakhstan and Azerbaijan as Post-Soviet Rentier States: Resource Incomes and Autocracy as a Double "Curse" in Post-Soviet Regimes', *Europe-Asia Studies*, 61(2): 109–40.

Friedberg, A. L. (2005) 'The Future of US-China Relations: Is Conflict Inevitable?' *International Security*, 30(2): 7–45.

Frieden, J. (2007) *Global Capitalism: Its Fall and Rise in the Twentieth Century*, New York: Norton.

Friedman, T. L. (2006) 'The First Law of Petropolitics', *Foreign Policy*, 154: 28–39.

Frynas, J. G. (2004) 'The Oil Boom in Equatorial Guinea', *African Affairs*, 103(413): 527–46.

Frynas, J. G. (2009) *Beyond Corporate Social Responsibility: Oil Multinationals and Social Challenges*, Cambridge: Cambridge University Press.

Frynas, J. G. and Paulo, M. (2007) 'A New Scramble for African Oil?: Historical, Political, and Business Perspectives', *African Affairs*, 106(423): 229–51.

Gajigo, O., Mutambatsere, E. and Ndiaye, G. (2012) 'Gold Mining in Africa: Maximizing Economic Returns for Countries', *Working Paper Series No. 147*, Tunis: African Development Bank Group.

Gallick, E. C. (1993) *Competition in the Natural Gas Pipeline Industry: An Economic Policy Analysis*, Westport, CT: Praeger.

Ganguly, S. (2007) 'The ONGC: Charting a New Course?' *Policy Brief*, The James A. Baker Institute for Public Policy, Rice University.

Gill, A. (1991) 'An Evaluation of Socially Responsive Planning in a New Resource Town', *Social Indicators Research*, 24: 177–204.

Gillespie, K. and Henry, C. M. (eds) (1995) *Oil in the New World Order*, Gainesville, FL: University Press of Florida.

Giordano, M., Giordano, M. and Wolf, A. (2005) 'International Resource Conflict and Mitigation', *Journal of Peace Research*, 42(1): 47–65.

Giurco, D., Prior, T., Mudd, G., Mason, L. and Behrisch, J. (2010) 'Peak Minerals in Australia: A Review of Changing Impacts and Benefits', *Cluster Research Report 1.2*, Institute for Sustainable Futures, University of Technology, Sydney.

Goldthau, A. and Witte J. M. (2009) 'Back to the Future or Forward to the Past?' *International Affairs*, 85(2): 373–90.

Goorha, P. (2006) 'The Political Economy of the Resource Curse in Russia', *Demokratizatsiya*, 14(4): 601–11.

Gordon, R. and Stenvoll, T. (2007) 'Statoil: A Study in Political Entrepreneurship', *Policy Brief*, The James A. Baker Institute for Public Policy, Rice University.

Graulau, J. (2008) '"Is Mining Good for Development?": The Intellectual History of an Unsettled Question', *Progress in Development Studies*, 8(2): 129–62.

Grayson, L. E. (1981) *National Oil Companies*, Chichester: John Wiley and Sons.

Gregory, C. (1980) *A Concise History of Mining*, Oxford: Pergamon.

Gunton, T. (2003) 'Natural Resources and Regional Development: An Assessment of Dependency and Comparative Advantage Paradigms', *Economic Geography*, 79(1): 67–94.

Guo, S. (2007) 'The Business Development of China's National Oil Companies: The Government to Business Relationship in China', *Policy Brief*, The James A. Baker Institute for Public Policy, Rice University.

Gurney, J. (1996) *Libya: The Political Economy of Energy*, Oxford: Oxford University Press, Oxford Institute of Energy Studies.

Haas, E. B. (1958) *Uniting Europe: Political, Social and Economic Forces*, London: Stevens & Sons.

Haber, S., Razo, A. and Maurer, N. (2003) *The Politics of Property Rights: Political Instability, Credible Commitments, and Economic Growth in Mexico, 1876–1929*, Cambridge: Cambridge University Press.

Haggard, S., MacIntrye, A. and Tiede, L. (2008) 'The Rule of Law and Economic Development', *Annual Review of Political Science*, 11: 205–34.

Halliday, F. (1981) *Soviet Policy in the Arc of Crisis*, London: Institute for Policy Studies.

Hanai, M. (1998) 'Formal and Garimpo Mining and the Environment in Brazil', in A. Warhurst (ed.), *Mining and the Environment*, Ottawa: IDRC.

Hardin, G. (1968) 'The Tragedy of the Commons', *Science*, 162: 1243–8.

Hart, A. and Jones, B. (2010/2011) 'How do Rising Powers Rise?' *Survival*, 52(6): 63–88.

Hartmen, H. and Mutmansky, J. (2002) *Introduction to Mining Engineering*, 2nd edn, New Jersey: John Willey.

Hartshorn, J. E. (1993) *Oil Trade: Politics and Prospects*, Cambridge: Cambridge University Press.

Harvey, D. (2003) *The New Imperialism*, Oxford: Oxford University Press.

Hassan, C., Olawoye, J. and Nnadozie, K. (2002) *Impact of International Trade and Multinational Corporations on the Environment and Sustainable Livelihoods of Rural Women in Akwa-Ibom State, Niger Delta Region, Nigeria*. Available at http://depot.gdnet.org/cms/conference/papers/4th_prl5.5.3_comfort_hassan_paper.pdf [Accessed 2 October 2012].

Haushofer, K. (2002) *Geopolitics of the Pacific Ocean*, New York: Edwin Mellen.

Hayes, M. H. (2004) *Algerian Gas to Europe: The Transmed Pipeline and Early Spanish Gas Import Projects*, Working Paper 27, Prepared for the Geopolitics of Natural Gas Study, a joint project of the Program on Energy and Sustainable Development at Stanford University and the James A. Baker III Institute for Public Policy of Rice University.

Helen, C., Roy, L. and Fiona S. (2002) 'People, Power, Participation: A Study of Mining – Community Relationships', *Minerals Mining and Sustainable Development*, International Association of Environment and Development.

Helm, D. (2011) 'Peak Oil and Energy Policy – A Critique', *Oxford Review of Economic Policy*, 27(1): 68–91.

Helms, L. (2008) 'Horizontal Drilling: North Dakota', *DMR Newsletter*, 35(1). Available at https://www.dmr.nd.gov/ndgs/Newsletter/NL0308/pdfs/Horizontal.pdf [Accessed 7 February 2013].

Henderson, J. (2012) *Rosneft – On the Road to Global NOC Status?* Oxford: Oxford Institute for Energy Studies.

Herbert, J. H. (1992) *Clean Cheap Heat: The Development of Residential Markets for Natural Gas in the United States*, New York: Praeger.

Hertzmark, D. (2007) 'Pertamina – Indonesia's State-Owned Oil Company', *Policy Brief*, The James A. Baker Institute for Public Policy, Rice University.

Hilson, G. and Murck, B. (2000) 'Sustainable Development in the Mining Industry: Clarifying the Corporate Perspective', *Resources Policy*, 26: 227–38.

Hinnebusch, R. (2003) *The International Politics of the Middle East*, Manchester: Manchester University Press.

Hirst, D. (1966) *Oil and Public Opinion in the Middle East*, Westport, CT: Praeger.

Hodge, R. A. (1995) *Assessing Progress Toward Sustainability: Development of a Systemic Framework and Reporting Structure*. Ph.D. dissertation, School of Urban Planning, Faculty of Engineering, McGill University, Montreal.

Hoff, K. and Stiglitz, J. E. (2005) 'The Creation of the Rule of Law and the Legitimacy of Property Rights: The Political and Economic Consequences of a Corrupt Privatization', *NBER Working Paper Series No. 11772*, Cambridge, MA: Cambridge University Press.

Horsnell, P. and Mabro, R. (1993) *Oil Markets and Prices: The Brent Market and the Formation of World Oil Prices*, Oxford: Oxford University Press, Oxford Institute for Energy Studies.

House of Lords. (1982) *Strategic Minerals*, 20th Report of the House of Lords Select Committee on the European Communities, Session 1981–82, HMSO.

Hubbert, M. K. (1956) 'Nuclear Energy and the Fossil Fuels', *Drilling and Production Practice*, 95: 1–40.

Hubbert, M. K. (1971) 'The Energy Resources of the Earth', *Scientific American*, 225(3): 1–61.

Hults, D. R. (2012a) 'Hybrid Governance: State Management of National Oil Companies', in D. G. Victor; D. R. Hults and M. C. Thurber (eds), *Oil and Governance: State-Owned Enterprises and the World Energy Supply*, Cambridge: Cambridge University Press.

Hults, D. R. (2012b) 'Petròleos de Venezuela, S. A. (PDVSA): From Independence to Subservience', in D. G. Victor, D. R. Hults and M. C. Thurber (eds), *Oil and Governance: State-Owned Enterprises and the World Energy Supply*, Cambridge: Cambridge University Press.

Humphreys, D. (2001) 'Sustainable Development: Can the Mining Industry Afford it?' *Resources Policy*, 27: 1–7.

Humphreys, D. (2009) 'Emerging Players in Global Mining', *Extractive Industries and Development Series, No. 5*, Washington, DC: World Bank.

Humphreys, D. (2010) 'The Great Metals Boom: A Retrospective', *Resources Policy*, 35(1): 1–13.

Humphreys, D. (2011) 'Emerging Miners and their Growing Competitiveness', *Mineral Economics*, 24(1): 7–14.

Humphreys, M., Sachs, J. D. and Stiglitz, J. E. (eds) (2007) *Escaping the Resource Curse*, New York: Columbia University Press.

IBRD/The World Bank. (1996) 'A Mining Strategy for Latin American and the Caribbean'. *World Bank Technical Paper No. 345*, Washington, DC: The World Bank.

IEA (International Energy Agency) (2008) *World Energy Outlook 2008*, Paris: OECD-IEA.

IEA (International Energy Agency) (2011) *World Energy Outlook 2011*, Paris.

IEA (International Energy Agency) (2012) 'FAQ's: Natural Gas'. Available at http://www.iea.org/aboutus/faqs/gas/ [Accessed 25 October 2012].

IEA-WEO (International Energy Agency – World Energy Outlook) (2011) 'Are we Entering a Golden Age of Gas?' *World Energy Outlook Special Report*. Available at http://www.worldenergyoutlook.org/media/weowebsite/2011/WEO2011_GoldenAgeofGasReport.pdf [Accessed 25 October 2012].

IEA-WEO (International Energy Agency – World Energy Outlook) (2012) 'Golden Rules for a Golden Age of Gas', *World Energy Outlook Special Report on Unconventional Gas*. Available at http://www.worldenergyoutlook.org/media/weowebsite/2012/goldenrules/WEO2012_GoldenRulesReport.pdf [Accessed 25 October 2012].

IFC (International Finance Corporation) Environment Division (2000) *Investing in People: Sustaining Communities Through Improved Business Practice*, Washington, DC: International Finance Corporation.

IFPEN (2012) 'Non-Conventional Hydrocarbons: Evolution or Revolution?' *Panorama*.

Innis, H. (1956) *Essays in Canadian Economic History*, Toronto, ON: Toronto University Press.

Ishiyama, J. (2002) 'Neopatrimonialism and the Prospects for Democratization in the Central Asian Republic', in Sally N. Cummings (ed.), *Power and Change in Central Asia*, London: Routledge.

Iyayi, F. (2006) 'Creating an Enabling Environment for Development in the Niger Delta: The Role of Labour', *SPDC (Shell Petroleum Development Company)*, Labour Seminar on Law, Order, Security and Sustainable Peace, Warri, Nigeria.

Jackson, R. T. (2002) 'Capacity Building in Papua New Guinea for Community Maintenance during and after Mine Closure', *Mining, Minerals and Sustainable Development No. 181*, London: International Institute for Environment and Development.

Jaffe, A. M. (2007) 'Iraq's Oil Sector: Past, Present and Future', *Policy Brief*, The James A. Baker Institute for Public Policy, Rice University.

Jaffe, A. M. and Elass, J. (2007) 'Saudi Aramco: National Flagship with Global Responsibilities', *Policy Brief*, The James A. Baker Institute for Public Policy, Rice University.

Jaffe, A. M. and Soligo, R. (2007) 'The International Oil Companies', *Policy Brief*, The James A. Baker Institute for Public Policy, Rice University.

Jenkins, H. (2004) 'Corporate Social Responsibility and the Mining Industry: Conflicts and Constructs', *Corporate Social Responsibility and Environmental Management*, 11(1): 23–34.

Johnston, D. (2007) 'How to Evaluate the Fiscal Terms of Oil Contracts', in M. Humphreys, J. D. Sachs and J. E. Stiglitz (eds), *Escaping the Resource Curse*, New York: Columbia University Press.

Julihn, C. E. (1928) *Summarized Data of Copper Production, (Economic Paper 1)*, Washington, DC: Bureau of Mines, US Department of Commerce, Government Printing Office.

Kaldor, M., Karl, Terry, L. and Said, Y. (eds) (2007) *Oil Wars*. London: Pluto Press.

Kapelus, P. (2002) 'Mining, Corporate Social Responsibility and the "Community": The Case of Rio Tinto, Richards Bay Minerals and the Mbonambi', *Journal of Business Ethics*, 39(3): 275–96.

Karasac, H. (2002) 'Actors of the New "Great Game": Caspian Oil Politics', *Journal of Southern Europe and the Balkans*, 4(1): 15–27.

Karl, T. (1997) *The Paradox of Plenty: Oil Booms and Petro-States*, Berkeley, CA: University of California Press.

Karl, T. (2000) 'Crude Calculations: OPEC Lessons for Caspian Leaders', in E. Robert and M. Rajan (eds) *Energy and Politics in Central Asia and the Caucasus*, New York: Rowman and Littlefield.

Karl, T. (2005) *Managing Iraq's Oil Wealth*, London: London School of Economics.

Karl, T. (2007) 'Oil-Led Development: Social, Political, and Economic Consequences', *CDDRL Working Papers*, Stanford University, Number 80, January.

Kathryn, F. -S. (1995) 'The Politics of Property Rights', *The American Political Science Review*, 89(4): 867–81.

Kathryn, F.-S. (1996) *The Transformation of Property Rights in the Gold Coast: An Empirical Analysis Applying Rational Choice Theory*, Cambridge: Cambridge University Press.

Katz, B. G. and Owen, J. (2009) 'Are Property Rights Enough? Re-Evaluating the Big-Bang Claim', *Economics of Transition*, 17(1): 75–96.

Kazeem, A. L. and Ademola, O. A. (2010) 'Pseudo-Component for Nigerian Heavy Oil and Bitumen', *SPE (Society of Petroleum Engineers)*, Nigerian Annual Conference and Exhibition, Tinapa-Calabar, Nigeria.

Keating, A. (2006) *Power, Politics and the Hidden History of Arabian Oil*, London: Saqi.

Keay, M. (2006) *Comment on the UK Energy Review and Decentralised Generation*, Oxford Institute for Energy Studies.

Kemp, D. (2003) 'Discovering Participatory Development through Corporate–NGO Collaboration: A Mining Industry Case Study', *CSRM Research Paper, 2*, University of Queensland: Centre for Social Responsibility in Mining. Available at http://www.csrm.uq.edu.au/docs/Paper2DKemp.pdf [Accessed 15 October 2011].

Keohane, R. O. (1984) *After Hegemony: Cooperation and Discord in the World Political Economy*, Princeton, NJ: Princeton University Press.

Keohane R. O. (ed.) (1986) *Neorealism and Its Critics*, New York: Columbia University Press.

Keohane, R. O. and Nye, J. (1977) *Power and Interdependence: World Politics in Transition*, Boston: Little and Brown.

Kim, Y. (2003) *The Resources Curse in a Post-Communist Regime: Russia Comparative Prospective*, Hants: Ashgate.

King, S. (2010) *Losing Control: The Emerging Threats to Western Prosperity*, New Haven, CT and London: Yale University Press.

Kissinger, H. (1979) *White House Years*, New York: Weidenfeld and Nicolson.

Kissinger, H. (1982) *Years of Upheaval*, New York: Weidenfeld and Nicolson.

Kitous, A., Criqui, P., Bellevrat, E. and Chateau, B. (2010) 'Transformation Patterns of the Worldwide Energy System—Scenarios for the Century with the POLES Model', *Energy Journal*, 31(1): 49–82.

Kitula, A. (2006) 'The Environmental and Socio-Economic Impacts of Mining on Local Livelihoods in Tanzania', *Journal of Cleaner Production*, 14(3–4): 405–14.

Klare, M. (2001) *Resource Wars*, New York: Henry Holt.

Klare, M. (2004) *Blood and Oil: The Dangers and Consequences of America's Growing Dependency on Imported Petroleum*, New York: Metropolitan Books.

Klare, M. (2008) *Rising Powers, Shrinking Planet: The New Geopolitics of Energy*, New York: Henry Holt.

Kobayashi, Y. (2007) 'Corporate Strategies of Saudi Aramco', *Policy Brief*, The James A. Baker Institute for Public Policy, Rice University.

Komesaroff, M. (2008) 'China Eyes Congo's Treasures', *Far Eastern Economic Review*, 171(3): 38–41.

Krapels, E. N. (1977) *Controlling Oil, British Oil Policy and the British National Oil Corporation*, Washington, DC: The United States Government Printing Office.

Kretzschmar, G. L., Kirchner, A. and Sharifzyanova, L. (2010) 'Resource Nationalism – Limits to Foreign Direct Investment', *The Energy Journal*, 31(2): 43–68.

Kuru, A. (2002) 'The Rentier State Model and Central Asian Studies: The Turkmen Case', *Alternatives: Turkish Journal of International Relations*, 1(1): 51–71.

Lachmann, R. (2000) *Capitalists in Spite of Themselves: Elite Conflict and Economic Transition in Early Modern Europe*, Oxford: Oxford University Press.

Laherrére, J. (2009) 'Peak Gold, Easier to Model than Peak Oil? – Part I', *The Oil Drum: Europe*. Available at http://europe.theoildrum.com/node/5989 [Accessed 2 October 2012].

Lanteigne, M. (2008) 'China's Maritime Security and the "Mallaca Dilemma" ', *Asian Studies*, 4(2): 143–61.

Laurence, D. (2011) 'Establishing a Sustainable Mining Operation: An Overview', *Journal of Cleaner Production*, 19(2–3): 278–84.

Le Billion, P. (2001) 'The Political Ecology of War: Natural Resources and Armed Conflicts', *Political Geography*, 20(5): 561–84.

Le Billion, P. (2004) 'The Geopolitical Economy of "Resource Wars" ', *Geopolitics*, 9(1): 1–28.

Le Billion, P. (2012) *Wars of Plunder: Conflicts, Profits and the Politics of Resources*, London: Hurst.

Levi, M. (1989) *Of Rule and Revenue*, Berkley, CA: University of California Press.

Lewis, S. W. (2007) 'Chinese NOCs and World Energy Markets: CNPC, Sinopec and CNOOC', *Policy Brief*, The James A. Baker Institute for Public Policy, Rice University.

Libecap, G. (1978) 'Economic Variables and the Development of Law: The Case of Western Mineral Rights', *Journal of Economic History*, 38(2): 338–62.

Libecap, G. (1996) 'Towards an Understanding of Property Rights', in L. J. Alston, T. Eggertsson and D. North (eds), *Empirical Studies in Institutional Change*, Cambridge: Cambridge University Press.

Lichbach, M. (June 2010) 'Charles Tilly's Problem Situations: From Class and Revolution to Mechanisms and Contentious Politics', *Perspectives on Politics*, 8(2): 543–9.

Lipschutz, R. D. (1989) *When Nations Clash: Raw Materials, Ideology and Foreign Policy*, New York: Harper and Row.

Lopez, L. (2012) 'Petronas: Reconciling Tensions between Company and State', in D. G. Victor, D. R. Hults and M. C. Thurber (eds), *Oil and Governance: State-Owned Enterprises and the World Energy Supply*, Cambridge: Cambridge University Press.

Losman, D. (2010) 'The Rentier State and National Oil Companies: An Economic and Political Perspective', *Middle East Journal*, 64(3): 427–45.

Lozano, L. (2009) 'Resource Nationalism in Latin America', *Mining Journal*, 16(Jan.).

Lucas, E. (2008) *The New Cold War: How the Kremlin Threatens Both Russia and the West*, London: Bloomsbury.

Luciani, G. (1987) 'Allocation vs. Production States: A Theoretical Framework', in H. Beblawi and G. Luciani (eds), *The Rentier State: Volume II*, London: Croom Helm.

Luciani, G. (1995) 'The Dynamics of Reintegration in the International Petroleum Industry', in K. Gillespie and C. M. Henry (eds), *Oil in the New World Order*, Gainesville, FL: University Press of Florida.

Luciani, G. (2001) 'National Oil Companies-Managers and Shareholders', *Oxford Energy Forum*, 46.

Luciani, G. (2004) 'NOC to IOC? National Oil Companies and International Oil Companies', *Oxford Energy Forum*, 57.

Luciani, G. and Salustri, M. (1998) 'Vertical Integration as a Strategy for Oil Security', in P. Stevens (ed.), *Strategic Positioning in the Oil Industry: Trends and Options*, London: I.B. Tauris.

Lujala, P. (2003) 'Classification of Natural Resources', *2003 ECPR Joint Session of Workshops*, Edinburgh.

Luning, S. (2011) 'Corporate Social Responsibility (CSR) for Exploration: Consultants, Companies and Communities in Processes of Engagements', *Resource Policy*, 37(2): 205–11.

Luong, P. J. (2000) 'The "Use and Abuse" of Russia's Energy Resources: Implications for State-Society Relations', in V. Sperling (ed.), *Building the Russian State: Institutional Crisis and the Quest for Democratic Governance*, Boulder, CO: Westview Press.

Luong, P. J. and Weinthal, E. (2010) *Oil Is Not a Curse: Ownership Structure and Institutions in Soviet Successor States*, Cambridge: Cambridge University Press.

Lynch, M. (1998) 'Crying Wolf: Warnings about Oil Supply', *M.I.T.*, March. Available at http://sepwww.stanford.edu/sep/jon/world-oil.dir/lynch/worldoil.html [Accessed 3 October 2012].

Lynch, M. (2002) *Mining in World History*, London: Reaktion Books.

Lynch, M. (2003) 'The New Pessimism about Petroleum Resources: Debunking the Hubbert Model (and Hubbert Modelers)', *Minerals and Energy*, 18(1): 21–32.

Lynch, M. (2009) 'Peak Oil is a Waste of Energy', *The New York Times*, August 25.

Mabro, R. (2005) 'The International Oil Price Regime: Origins, Rationale and Assessment', *Journal of Energy Literature*, 11(1): 3–20.

Mabro, R. (2006a) 'The Peak Oil Theory', *Oxford Energy Comment*, September.

Mabro, R. (ed.) (2006b) *Oil in the 21st Century: Issues, Challenges and Opportunities*, Oxford: Oxford University Press, Organization of the Petroleum Exporting Countries.

Mabro, R. (2010) 'On Oil Peak or Peaks', *Oxford Energy Journal*, 80: 9–11.

Mackinder, H. (1919) *Democratic Ideals and Reality: A Study in the Politics of Reconstruction*, London: Constable.

Mackintosh, W. (1964) *The Economic Background of Dominion-Provincial Relations*, Toronto, ON: McClelland and Stewart.

Madelin, H. (1974) *Oil and Politics*, London: Saxon House/Lexington Books.

Mahan, A. T. (1890) *The Influence of Sea Power upon History*, New York: Hill and Wang.

Mahdavy, H. (1970) 'The Patters and Problems of Economic Development in Rentier States: The Case of Iran', in M. A. Cook (ed.), *Studies in The Economic History of The Middle East*, London: Oxford University Press.

Mallorquín, C. (2007) 'Celso Furtado and Development: An Outline', *Development in Practice*, 17(6): 807–19.

Marcel, V. with Mitchell, J. V. (2006) *Oil Titans: National Oil Companies in the Middle East*, Washington, DC: Brookings Institution.

Mares, D. and Altamirano, N. (2007) 'Venezuela's PDVSA and World Energy Markets', *Policy Brief*, The James A. Baker Institute for Public Policy, Rice University.

Masset, J. -M. (avec les contributions de P-R Bauquis, D. Gautier & A. Perrodon) (2009) 'Pétrole, Gaz: pic ou plateau', *Revue de l'énergie*, 591: 318–25.

Maull, H. W. (1986) 'South Africa's Minerals: The Achilles Heel of Western Economic Security?' *International Affairs*, 62(4): 619–26.

Maurice, C. and Smithson, C. W. (1984) *The Doomsday Myth, 10,000 Years of Economic Crises*, Stanford, CA: Hoover Institution Press.

May, D., Prior, T., Cordell, D. and Giurco, D. (2011) 'Peak Minerals: Theoretical Foundations and Practical Applications', *Natural Resources Research*, 21(1): 23–60.

McAllister, M., Scoble, M. and Veiga, M. (2001) 'Mining with Communities', *Natural Resources Forum*, 25(3): 191–202.

McHarg, B. B., Bradbrook, A. and Godden, L. (eds) (2010) *Property and the Law in Energy and Natural Resources*, Oxford: Oxford University Press.

McPherson, C. (2009) 'State Participation in the Natural Resources Sectors: Evolution, Issues and Outlook', in P. Daniel, M. Keen and Ch. McPherson (eds), *The Taxation of Petroleum and Minerals: Principles, Problems and Practice*, London: Routledge.

McPherson, C. and MacSearraigh, S. (2007) 'Corruption in the Petroleum Sector', in J. E. Campos and S. Pradhan (eds), *The Many Faces of Corruption: Tracking Vulnerabilities at the Sector Level*, Washington, DC: The World Bank.

Meadows, D. H., Meadows, D. L., Randers, J. and Behrens, W. W. (1974) *The Limits to Growth*, London: Pan Books.

Melcher, F., Sitnikova, M., Graupner, T., Martin, N., Oberthür, T., Henjes-Kunst, F., Gäbler, E., Gerdes, A., Brätz, H., Davis, D. and Dewaele, S. (2008) 'Fingerprinting of Conflict Minerals: Columbite-Tantalite ("Coltan") Ores', *SGA News (Society for Geology Applied to Mineral Deposits)*, 23(June). Available at http://e-sga.org/fileadmin/sga/newsletter/news23/SGANews23.pdf [Accessed 31 May 2012].

Meyer, R., Attanasi, E. and Freeman, P. (2007) 'Heavy Oil and Natural Bitumen Resources in Geological Basins of the World', *US Geological Survey Open-File Report 2007-1084*. Available at http://pubs.usgs.gov/of/2007/1084/ [Accessed 10 July 2009].

Miller, J. A., Fine, D. I. and McMichael, R. D. (eds) (1980) *The Resource War in 3-D: Dependency, Diplomacy, and Defense*, Pittsburgh, PA: World Council on Foreign Affairs.

Milmo, C. (2010) 'Concern as China Clamps Down on Rare Earth Exports', *The Independent*, 2(Jan.).

Mineral Statistics of the United Kingdom (1996) 'Cornish Mineral Production'. Available at http://projects.exeter.ac.uk/mhn/MS/co-intro.html [Accessed 15 October 2011].

Miranda, M., Burris, P., Binchang, J. F., Shearman, P., Briones, J. O., La Vina, A. and Menard, S. (2003) *Mining and Critical Ecosystems: Mapping the Risks*, Washington, DC: World Resources Institute.

Mitchell, T. (2009) 'Carbon Democracy', *Economy and Society*, 38(3): 399–432.

Mitchell, T. (2011) *Carbon Democracy: Political Power in the Age of Oil*, London: Verso.

MMSD (Mining, Minerals and Sustainable Development) (2002) *Breaking New Ground: A Report of the Mining, Minerals and Sustainable Development Project*, London: Earthscan.

Mommer, B. (1998) *The New Governance of Venezuelan Oil*, Oxford: Oxford Institute for Energy Studies.

Mommer, B. (2000) *The Governance of International Oil: The Changing Rules of the Game*, Oxford: Oxford Institute for Energy Studies.

Mommer, B. (2001) *Fiscal Regimes and Oil Revenues in the UK, Alaska and Venezuela*, Oxford: Oxford Institute for Energy Studies.

Mommer, B. (2002) *Global Oil and the Nation State*, Oxford: Oxford University Press.

Monopolies and Mergers Commission (UK). (1988) *The Government of Kuwait and the British Petroleum Company – A Report on the Merger Situation*, London: Her Majesty's Stationary Office.

Montgomery, C. T. and Smith, M. B. (2010) 'History of an Enduring Technology', *JPT*, December: 26–41. Available at http://www.spe.org/jpt/print/archives/2010/12/10Hydraulic.pdf [Accessed 7 February 2013].

Moore, P. (2001) 'What Makes Successful Business Lobbies? Business Associations and the Rentier State in Jordan and Kuwait', *Comparative Politics*, 33(2): 127–47.

Morgenthau, H. J. (1960) *Politics Among Nations: The Struggle for Power and Peace*, New York: Knopf.

Morris, D. (2006) 'The Chance to go Deep: US Energy Interests in West Africa', *American Foreign Policy Interests*, 28(3): 225–38.

Müller, M. (2010) 'Revenue Transparency to Mitigate the Resource Curse in the Niger Delta? Potential and Reality of NEITI', *Occasional Paper 5*, Bonn International Centre for Conversion (BICC) GmbH.

Nanto, D. K., Jackson, J. K., Morrison, W. M. and Kumins, L. (2005) *China and the CNOOC Bid for Unocal: Issues for Congress*, Washington, DC: Congressional Research Service.

National Research Council (2008) *Minerals, Critical Minerals, and the US Economy*, Washington, DC: The National Academies Press.

NaturalGas.org (2012) 'Unconventional Natural Gas Resources'. Available at http://www.naturalgas.org/overview/unconvent_ng_resource.asp [Accessed 25 October 2012].

Natural Gas Europe (2012) 'Poland Positions as European Shale Gas Pioneer as Tax Regulation Framework Unveiled', 18(October). Available at http://www.naturalgaseurope.com/poland-tax-regulation-framework-shale-gas [Accessed 25 October 2012].

Natural Resource Canada (2011) 'Natural Gas: A Primer'. Available at http://www.nrcan.gc.ca/energy/sources/natural-gas/1233#conventional [Accessed 25 October 2012].

NETL (National Energy Technology Laboratory). (2011) 'Shale Gas: Applying Technology to Solve America's Energy Challenges'. Available at http://www.netl.doe.gov/technologies/oil-gas/publications/brochures/Shale_Gas_March_2011.pdf [Accessed 25 October 2012].

Niamh, K. (ed.) (2010a) *Energy Fundamentals: Understanding the Oil and Gas Industries*, New York: Energy Intelligence Group, Incorporated.

Niamh, K (ed.) (2010b) *World LNG Outlook 2010–2011*, New York: Energy Intelligence Group, Incorporated.

Nicita, A. and Pagano, U. (2008) 'Law and Economics in Retrospect', in E. Brosseau and J. -M. Glachant (eds), *New Institutional Economics. A Guidebook*, Cambridge: Cambridge University Press.

Niebuhr, R. (1960) *Moral Man and Immoral Society: A Study in Ethics and Politics*, New York: Charles Scribner.

Nolan, P. A. and Thurber, M. C. (2012) 'On the State's Choice of Oil Company: Risk Management and the Frontier of the Petroleum Industry', in D. G. Victor, D. R. Hults and M. C. Thurber (eds), *Oil and Governance: State-Owned Enterprises and the World Energy Supply*, Cambridge: Cambridge University Press.

Noreng, O. (1980) *The Oil Industry and Government Strategy in the North Sea*, London: Croom Helm.

Noreng, O. (1997) *Oil and Islam: Social and Economic Issues*, Chichester: John Wiley and Sons.

Noreng, O. (2002) *Crude Power: Politics and the Oil Market*, London: I.B. Tauris.

Noronha, L. (2001) 'Designing Tools to Track Health and Well-Being in Mining Regions of India', *Natural Resource Forum*, 25: 53–65.

North, D. C. (1990) *Institutions, Institutional Change and Economic Performance*, Cambridge: Cambridge University Press.

North, D. C., Wallis, J. J. and Weingast, B. R. (2009) *Violence and Social Order: A Conceptual Framework for Interpreting Recorded Human History*, New York: Cambridge University Press.

Nwokeji, U. (2007) 'The Nigerian National Petroleum Corporation and the Development of the Nigerian Oil and Gas Industry: History, Strategies and Current Directions', *Policy Brief*, The James A. Baker Institute for Public Policy, Rice University.

Obi, C. (2010) 'The Petroleum Industry: A Paradox or (Sp)oiler of Development?' *Journal of Contemporary African Studies*, 28(4): 443–57.

Ochola, S. A. (1975) *Minerals in African Underdevelopment*, London: Bogle-L'Ouverture.

Odell, P. (2003) 'The Global Energy Outlook for the 21st Century', *Annual Lecture NOGEPA and the Oranje-Nassau Groep*, Wassenaar.

Odell, P. (2010) 'Why we do not have to Worry about Peak Oil?' *European Energy Review*, January 15.

O'Hagan, S. and Cecil, B. (2007) 'A Macro-Level Approach to Examining Canada's Primary Industry Towns in a Knowledge Economy', *Journal of Rural and Community Development*, 2: 18–43.

Ojo, G. U. (2012) 'Economic Diversification and Second-Tier Political Conflict: assessing Bitumen Political Ecologies in Southwest Nigeria', *Singapore Journal of Tropical Geography*, 33(1): 49–62.

Okruhlik, G. (1999) 'Rentier Wealth, Unruly Law, and the Rise of Opposition: The Political Economy of Oil States', *Comparative Politics*, 31(3): 295–315.

Olcott, M. B. (2007) 'Kazmunaigaz: Kazakhstan's National Oil and Gas Company', *Policy Brief*, The James A. Baker Institute for Public Policy, Rice University.

Olson, M. (2000) *Power and Prosperity: Outgrowing Communist and Capitalist Dictatorships*, New York: Basic Books.

Omorogbe, Y. and Oniemoal, P. (2010) 'Property Rights in Oil and Gas under Domanial Regimes', in A. McHarg, B. Barton, A. Bradbrook and L. Godden (eds), *Property and the Law in Energy and Natural Resources*, Oxford: Oxford University Press.

Onorato, W. T. (1995) 'Legislative Frameworks used to Foster Petroleum Development', *Policy Research Working Paper No. 1420*, Washington, DC: World Bank.

Onyeukwu, A. (2007) 'Resource Curse in Nigeria: Perception and Challenges', *International Policy Fellowship Program 2006/2007*, Central European University, Centre for Policy Studies.

Orogun, P. S. (2010) 'Resource Control, Revenue Allocation and Petroleum Politics in Nigeria: The Niger Delta Question', *GeoJournal*, 75(5): 459–507.

Östensson, O. (1997) 'A Brief Background on Social Issues and Mining', *Asian/Pacific Workshop on Managing the Social Impacts of Mining*, UNCTAD (United Nations Conference on Trade and Development), Bandung, Indonesia.

Ostrom, E. (2005) *Understanding Institutional Diversity*, Princeton, NJ: Princeton University Press.

Ostrom, E., Dietz, T., Dolšak, N., Stern, P. C. and Stovich, S. (eds) (2003) *The Drama of the Commons*, Washington, DC: National Academy Press.

Ostrowski, W. (2010) 'Comparative Politics of Energy and Minerals: Concepts, Debates and Gaps', *Polinares Working Paper, no. 9*, September.

Ostrowski, W. (2011a) 'Rentierism, Dependency and Sovereignty in Central Asia', in S. N. Cummings and R. Hinnebusch (eds), *Sovereignty After Empire: Comparing the Middle East and Central Asia*, Edinburgh: Edinburgh University Press.

Ostrowski, W. (2011b) *Politics and Oil in Kazakhstan*, London: Routledge.

Otto, J. (1996) 'The Changing Regulatory Framework for Mining Ventures', *Journal of Energy and Natural Resources Law*, 14(3): 251–61.

Parra, F. (2004) *Oil Politics: A Modern History of Petroleum*, London: I.B.Tauris.

Parsons, T. (1951) *The Social System*, Glencoe, IL: Free Press.

Pearson, P. (2009) 'Climate Change, Energy and Innovation: A UK Historical Perspective', *papers presented at the Finance, Food and Energy Crises (Symposium for SCOPE)*, Imperial College, London.

Peebles, M. W. H. (1980) *Evolution of the Gas Industry*, London and Basingstoke: Macmillan.

Pehrson, E. W. (1929) *Summarized Data of Zinc Production (Economic Paper 2)*, Washington, DC: Bureau of Mines, US Department of Commerce, Government Printing Office.

Penrose, E. and Penrose, E. F. (1978) *Iraq: International Relations and National Development*, London: Benn.

Penrose, E. T. (1968–) *The Large International Firm in Developing Countries: The International Petroleum Industry*, London: George Allen and Unwin.

Pesaran, M. H. and Samiei, H. (1995) 'Forecasting Ultimate Resource Recovery', *International Journal of Forecasting*, 11(4): 543–55.

Petras, J. (1977) 'State Capitalism and the Third World', *Development and Change*, 8(1): 1–17.

Pirani, S., Stern, J. and Yafimava, K. (February 2009) *The Russo-Ukrainian Gas Dispute of 2009: A Comprehensive Assessment*, Oxford Institute for Energy Studies Working Paper.

Pistor, K. (2002) 'The Standardization of Law and its Effect on Developing Countries', *American Journal of Comparative Law*, 97: 97–130.

Platts (2012) 'China Targets 6.5 Bcm Shale Gas Output by 2015 under Five-Year Plan', 16(Mar). Available at http://www.platts.com/RSSFeedDetailedNews/RSSFeed/NaturalGas/8069251 [Accessed 25 October 2012].

Popper, K. (1957) *The Poverty of Historicism*, New York: Harper.

Porter, E. D. (1995) 'Are we Running out of Oil?' *Discussion Paper #081*, American Petroleum Institute.

Porter, M. and Kramer, M. (2006) 'Strategy and Society: The Link between Competitive Advantage and Corporate Social Responsibility', *Harvard Business Review*, 84(12): 78–93.

Poussenkova, N. (2007) 'Lord of the Rigs: Rosneft as a Mirror of Russia's Evolution', *Policy Brief*, The James A. Baker Institute for Public Policy, Rice University.

Prebisch, R. (1950) *Economic Development of Latin America and Its Principal Problems*, New York: United Nations.

Radetzki, M. (1985) *State Mineral Enterprises*, Washington, DC: Resources for the Future.

Radetzki, M. (2002) 'Is Resource Depletion a Threat to Human Progress? Oil and other Critical Exhaustible Materials', *Energy Sustainable Development-A Challenge for the New Century, Plenary Papers Presented at the 9th International Energy Conference*, Mineral and Energy Economy Research Institute of the Polish Academy of Sciences and the International Energy Foundation, Krakow, Poland.

Radetzki, M. (2009) 'Seven Thousand Years in the Service of Humanity – The History of Copper, the Red Metal', *Resources Policy*, 34: 176–84.

Radetzki, M. (2010) 'Peak Oil and other Threatening Peaks – Chimeras without Substance', *Energy Policy*, 38: 6566–9.

Reno, W. (2011) *Warfare in Independent Africa*, Cambridge: Cambridge University Press.

RFF (Resources for the Future) (2012) 'A Review of Shale Gas Regulations by State', *Center for Energy Economics and Policy*. Available at http://www.rff.org/centers/energy_economics_and_policy/Pages/Shale_Maps.aspx [Accessed 25 October 2012].

Rhodes, Robert (ed.) (1974) *Winston S. Churchill: His Complete Speeches, 1897–1963*, New York: James.

Ritter, A. R. (2000) 'From Fly-in, Fly-out to Mining Metropolis', in M. Gary and R. Felix (eds), *Large Mines and the Community: Socioeconomic and Environmental Effects in Latin America, Canada, and Spain*, Washington, DC: World Bank.

Robinson, C. (1993) *Energy Policy: Errors, Illusions and Market Realities: Occasional Paper 90*, London: The Institute of Economic Affairs.

Rooney, R. C., Bayley, S. E. and Schindler, D. W. (2012) 'Oil Sands Mining and Reclamation Cause Massive Loss of Peatland and Stored Carbon', *PNAS*, 109(13): 4933–7.

Ross, M. (1999) 'The Political Economy of the Resource Curse', 51(2): 297–322.

Ross, M. (2001) 'Does Oil Hinder Democracy?' *World Politics*, 53(3): 325–61.

Ross, M. (2003) 'Nigeria's Oil Sector and the Poor', *Nigeria: Drivers for Change*, UK DFID (Department for International Development).

Ross, M. (2004) 'What we Know about Natural Resources and Civil War', *Journal of Peace Research*, 41(3): 337–56.

Ross, M. (2006) 'A Closer Look at Oil, Diamonds, and Civil War', *Annual Review of Political Science*, 9: 265–300.

Ross, M. (2012) *The Oil Curse: How Petroleum Wealth Shapes the Development of Nations*, Princeton, NJ: Princeton University Press.

Rosser, A. (2006) 'The Political Economy of the Resource Curse: A Literature Survey', *IDS Working Paper* 268.

Rostow, W. (1960) *The Stages of Economic Growth: A Non-Communist Manifesto*, Cambridge: Cambridge University Press.

Roxborough, I. (1988) 'Modernization Theory Revisited: A Review Article', *Comparative Studies in Society and History*, 30(4): 753–61.

Rudiger, A. (2005) 'Can Russia Break the "Resource Curse"?' *Eurasian Geography and Economics*, 46(8): 584–609.

Ruhrgas (1991) *Gas Today and Tomorrow*, Essen: Ruhrgas.

Russett, B. M. (1993) *Grasping the Democratic Peace: Principles for a Post-Cold War World*, Princeton, NJ: Princeton University Press.

Rutledge, I. and Wright, P. (eds) (2011) *UK Energy Policy and the End of Market Fundamentalism*, Oxford Institute for Energy Studies.

Sachs, J. and Warner, A. (1995) 'Economic Convergence and Economic Policies', *Harvard Institute of Economic Research*, Working Papers 1715, Harvard – Institute of Economic Research.

Salim, E. (2003) *Extractive Industries Review, Striking a Better Balance*. Washington, DC: International Finance Corporation and The World Bank. Available at http://web.worldbank.org/WBSITE/EXTERNAL/TOPICS/EXTOGMC/0,,contentMDK:2030 6686~menuPK:592071~pagePK:148956~piPK:216618~theSitePK:336930,00.html [Accessed 2 October 2012].

Sampson, A. (1975) *The Seven Sisters: The Great Oil Companies and the World They Made*, London: Hodder and Stoughton.

Schafrik, S. and Kazakidis, V. (2011) 'Due Diligence in Mine Feasibility Studies for the Assessment of Social Risk', *International Journal of Mining, Reclamation and Environment*, 25(1): 86–101.

Schwarz, R. (2008) 'The Political Economy of State-Formation in the Arab Middle East: Rentier States, Economic Reform, and Democratization', *Review of International Political Economy*, 15(4): 599–621.

Scott, A. (2008) *The Evolution of Resource Property Rights*, Oxford: Oxford University Press.

Seymour, I. (1980) *OPEC: Instrument of Change*, London: Macmillan.

Shaffer, B. (2009) *Energy Politics*, Philadelphia, PA: University of Pennsylvania Press.

Shambayati, H. (1994) 'The Rentier State, Interest Groups, and the Paradox of Autonomy: State and Business in Turkey and Iran', *Comparative Politics*, 26(3): 307–31.

Shaxson, N. (2007) *Poisoned Wells: The Dirty Politics of African Oil*, London: Palgrave Macmillan.

Shaxson, N. (2009) 'Nigeria's Extractive Industries Transparency Initiative: Just a Glorious Audit?' *Chatham House Programme Paper*.

Shell (2008) *Shell Energy Scenarios to 2050*, Shell International BV.

Shwardan, B. (1985) *The Middle East, Oil and the Great Powers*, Boulder, CO: Westview Press.

Silverstein, K. (2009) 'Invisible Hands: The Secret World of the Oil Fixer', *Harpers Magazine*, March.

Skeet, I. (1988) *OPEC: Twenty-Five Years of Prices and Politics*, Cambridge: Cambridge University Press.

Skocpol, T. (1985) 'Bringing the State Back in: Strategies of Analysis in Current Research', in P. Evans, D. Rueschemeyer and T. Skocpol (eds), *Bringing the State Back In*, Cambridge: Cambridge University Press.

Smith, L. A. (1929) *Summarized Data of Lead Production (Economic Paper 5)*, Washington, DC: Bureau of Mines, US Department of Commerce, Government Printing Office.

Snyder, R. (2006) 'Does Lootable Wealth Breed Disorder? A Political Economy of Extraction Framework', *Comparative Political Studies*, 399(8): 943–68.

Soares de Oliveira, R. (2007a) *Oil and Politics in the Gulf of Guinea*, New York: Columbia University Press.

Soares de Oliveira, R. (2007b) 'Business Success, Angola-Style: Postcolonial Politics and the Rise and Rise of Sonangol', *Journal of Modern African Studies*, 45(4): 595–619.

Söderholm, P. (2003) 'Reflections on the Oil Depletion Controversy', *Minerals and Energy*, 18(2): 2–6.

Šolar, S. V., Shields, D. J. and Miller, M. D. (2009) 'Mineral Policy in the Era of Sustainable Development: Historical Context and Future Content', *RMZ – Materials and Geoenvironment*, 56(3): 304–21.

Southalan, J. (2011) 'What are the Implications of Human Rights for Minerals Taxation?' *Resources Policy*, 36(3): 214–26.

Southalan, J., Culotta, K. and Fallon, A. (2011) 'Indigenous People and Resource Development: A Rapidly Changing Legal Landscape', *OGEL Journal (Oil, Gas and Energy Law Intelligence)*, 4.

Sprout, H. and Sprout, M. (1971) *Towards a Politics of Planet Earth*, New York: Van Nostrand Reinhold.

Stern, J. (2005) *The Future of Russian Gas and Gazprom*, Oxford: Oxford University Press.

Stern, J. (ed.) (2008) *Natural Gas in Asia: The Challenges of Growth in China, India, Japan and Korea*, 2nd edn, Oxford: Oxford University Press.

Stern, J. P. (2005) *The Future of Russian Gas and Gazprom*, Oxford: Oxford University Press.

Stevens, P. (1975) *Joint Venture in Middle East Oil 1957–75*, Beirut: Middle East Economic Consultants.

Stevens, P. (ed.) (1998) *Strategic Positioning in the Oil Industry: Trends and Options*, London: I.B. Tauris.

Stevens, P. (2003a) 'Resource Impact: Curse or Blessing? A Literature Survey', *Journal of Energy Literature*, 9(1): 3–42.

Stevens, P. (2003b) 'National Oil Companies: Good or Bad? A Literature Survey', *World Bank Workshop*, Washington, DC.

Stevens, P. (2008a) 'National Oil Companies and International Oil Companies in the Middle East: Under the Shadow of Government and the Resource Nationalism Cycle', *Journal of World Energy Law & Business*, 1(1): 5–30.

Stevens, P. (2008b) 'Oil Wars: Resource Nationalism and the Middle East', in P. Andrews-Speed (ed.), *International Competition for Resources: The Role of Law, the State and of Markets*, Dundee: University Press Dundee.

Stevens, P. (2008c) 'The Coming Oil Supply Crunch', *A Chatham House Report*, London: Chatham House.

Stevens, P. (2008d) 'A Methodology for Assessing the Performance of National Oil Companies', *The World Bank (Oil, Gas and Mining Policy Division)*.

Stevens, P. (2009) *The Coming Oil Supply Crunch*, Chatham House Report, London.

Stevens, P. (2011) 'Cooperation between Producers and Consumers', in Robert E. Looney (ed.), *A Handbook of Oil Politics*, London: Routledge.

Stevens, P. (2012a) 'An Embargo on Iranian Crude Oil Exports: How Likely and with what Effect?' *Chatham House Programme Paper*, London: Chatham House.

Stevens, P. (2012b) 'The Arab Uprisings and the International Oil Markets', *Chatham House Briefing Paper*, London: Chatham House.

Stevens, P. (2012c) 'Saudi Aramco: The Jewel in the Crown', in D. G. Victor, D. R. Hults and M. C. Thurber (eds), *Oil and Governance: State-Owned Enterprises and the World Energy Supply*, Cambridge: Cambridge University Press.

Stevens, P. and Dietsche, E. (2008) 'Resource Curse: An Analysis of Causes, Experiences and Possible Ways Forward', *Energy Policy*, 36(1): 56–65.

Stiglitz, J. E. (2007) 'What is the Role of the State?' in M. Humphreys, J. D. Sachs and J. E. Stiglitz (eds), *Escaping the Resource Curse*, New York: Columbia University Press.

Stocking, G. W. (1970) *Middle East Oil*, Nashville, TN: Vanderbilt University Press.

Stojanovski, O. (2012) 'Handcuffed: An Assessment of Pemex's Performance and Strategy', in D. G. Victor, D. R. Hults and M. C. Thurber (eds), *Oil and Governance: State-Owned Enterprises and the World Energy Supply*, Cambridge: Cambridge University Press.

Stokes, D. and Raphael, S. (2010) *Global Energy Security and American Hegemony*, Baltimore, MD: Johns Hopkins Press.

Sumi, L. and Thomsen, S. (2001) *Mining in Remote Areas, Issues and Impacts: A Community Primer*, Ontario: MiningWatch Canada/Mines Alert.

Taylor, I. (2006) 'China's Oil Diplomacy in Africa', *International Affairs*, 82(5): 937–59.

Thurber, M. C. and Istad, B. T. (2012) 'Norway's Evolving Champion: Statoil and the Politics of State Enterprise', in D. G. Victor, D. R. Hults and M. C. Thurber (eds), *Oil and Governance: State-Owned Enterprises and the World Energy Supply*, Cambridge: Cambridge University Press.

Tilly, C. (1985) 'War Making and State Making as Organized Crime', in P. Evans, D. Rueschemeyer and T. Skocpol (eds), *Bringing the State Back In*, Cambridge: Cambridge University Press.

Tilly, C. (1992) *Coercion, Capital and European States: AD 990–1992*, Oxford: Blackwell.

Tilton, J. E. (2003) 'Assessing the Threat of Mineral Depletion', *Minerals and Energy*, 18(1): 33–42.

Tompson, W. (2006) 'A Frozen Venezuela? The "Resource Curse" and Russian Politics', in M. Ellman (ed.), *Russia's Oil and Natural Gas: Bonanza or Curse?* London: Anthem.

Topping, P. (2011) *Pre-industrial Mines and Quarries: Introduction to Heritage Assets*, English Heritage.

Tordo, S. with Johnston, D. and Johnston, D. (2010) 'Petroleum Exploration and Production Rights: Allocation Strategies and Design Issues', *World Bank Working Paper No. 179*, The World Bank.

Tordo, S. (2007) 'Fiscal Systems for Hydrocarbons: Design Issues', *World Bank Working Paper No. 123*, The World Bank.

Tordo, S. (2011) *National Oil Companies and Value Creation*, Washington, DC: World Bank.

Townsend, P. K. and Townsend, W. H. (2004) *Assessing an Assessment: The Ok Tedi Mine Conference Proceedings. Bridging Scales and Epistemologies: Linking Local Knowledge and Global Science in Multi-Scale Assessments*. Available at http://www.unep.org/maweb/documents/bridging/papers/townsend.patricia.pdf [Accessed 8 February 2013].

Toye, J. and Toye, R. (2003) 'The Origins and Interpretation of the Prebisch-Singer Thesis', *History of Political Economy*, 35(3): 437–67.

Treisman, D. (2010) 'Rethinking Russia: Is Russia Cursed by Oil?' *Journal of International Affairs*, 63(2): 85–102.

Turner, L. (1978) *Oil Companies in the International System*, London: Royal Institute of International Affairs.

Twerefou, D. K. (2009) 'Mineral Exploitation, Environmental Sustainability and Sustainable Development in EAC, SADC and ECOWAS Regions', *African Trade and Policy Centre Work in Progress 79*, Economic Commission for Africa.

UK Energy Research Centre. (2009) *Global Oil Depletion: An Assessment of the Evidence for a Near Term Peak in Global Oil Production*, UKERC. Available at http://www.ukerc.ac.uk/support/Global%20Oil%20Depletion [Accessed 3 October 2012].

Umbeck, J. (1977) 'The California Gold Rush: A Study of Emerging Property Rights', *Explorations in Economic History*, 14: 197–226.

Umbeck, J. (1981) *A Theory of Property Rights with Application to the California Gold Rush*, Ames, IA: Iowa State University Press.

Umhau, J. B. (1932) *Summarized Data of Tin Production (Economic Paper 13)*, Washington, DC: Bureau of Mines, US Department of Commerce, Government Printing Office.

UNCRET (United Nations Centre for Natural Resources, Energy and Transport) (1980) *State Petroleum Enterprises in Developing Countries*, New York: Pergamon Press.

UNCTAD (2007) *World Investment Report: Part II – Transnational Corporations, Extractive Industries and Development*, Geneva: United Nations Conference on Trade and Development.

UNDP (United Nations Development Programme) (2006) *Niger Delta Human Development Report*, Lagos: Perfect Printer.

UNECA (United Nations Economic Commission for Africa) (2011) *Minerals and Africa's Development: The International Study Group Report on Africa's Mineral Regimes*, Addis Ababa: Economic Commission for Africa and African Union.

UNEP (United Nations Environmental Programme) (2011) *Environmental Assessment of Ogoniland*, Nairobi: United Nations Environment Programme.

United Nations (2011) *Guiding Principles on Business and Human Rights: Implementing the United Nations' 'Respect, Protect and Remedy' Framework*, HR/PUB/11/4, New York: United Nations Human Rights Office of the High Commissioner.

United Nations Conference on Trade and Development (UNCTAD) (2007) *World Investment Report: Transnational Corporations, Extractive Industries and Development*, Geneva: United Nations Conference on Trade and Development.

United Nations Environment Program (UNEP) (2009) 'From Conflict to PeaceBuilding: The Role of Natural Resources and the Environment', *Report: United Nations Environment Program*, February.

US Government (2005) 'Energy Policy Act of 2005'. Available at http://www.fedcenter.gov/Documents/index.cfm?id=2969 [Accessed 25 October 2012].

US House of Representative (2011) 'Chemicals used in Hydraulic Fracturing', *United States House of Representatives Committee on Energy and Commerce Minority Staff*, April. Available at http://democrats.energycommerce.house.gov/sites/default/files/documents/Hydraulic%20Fracturing%20Report%204.18.11.pdf [Accessed 25 October 2012].

United States Federal Trade Commission (1952) *The International Petroleum Cartel; Staff Report to [i.e. of] the Federal Trade Commission Submitted to the Subcommittee on Monopoly of the Select Committee on Small Business, United States Senate*, Washington, DC, US Govt. Print. Off.

US Senate, Subcommittee on Multinational Corporations of the Committee on Foreign Relations (1974) *The International Petroleum Cartel, the Iranian Consortium and US National Security*, Washington, DC, US Govt. Print. Off.

Valenzuela, S. and Valenzuela, A. (1978) 'Modernisation and Dependency: Alternative Perspective in the Study of Latin American Underdevelopment', *Comparative Politics*, 10(4): 535–57.

Van Groenendaal, W. (1999) *The Economic Appraisal of Natural Gas Projects (Oxford Institute for Energy Studies)*, Oxford: Oxford University Press.

Van den Linde, C. (1991) *Dynamic International Oil Markets*, London: Kluwer.

Van den Linde, C. (2000) *The State and the International Oil Market: Competition and the Changing Ownership of Crude Oil Assets*, London: Kluwer.

Veiga, M. M., Scoble, M. and McAllister, M. L. (2001) 'Mining with Communities', *Natural Resources Forum*, 25: 191–202.

Vernon, R. (1971) *Sovereignty at Bay: The Multinational Spread of US Enterprises*, New York: Basic Books.

Vernon, R. (1985) 'Sovereignty at Bay: Ten Years after', in T. H Moran (ed.), *Multinational Corporations: The Political Economy of Foreign Direct Investment*, Lexington, MA: Lexington Books.

Victor, D. G., Hults, D. R. and Thurber, M. (eds) (2012) *Oil and Governance: State-Owned Enterprises and the World Energy Supply*, Cambridge: Cambridge University Press.

Victor, N. and Sayfer, I. (2012) 'Gazprom: The Struggle for Power', in D. G. Victor, D. R. Hults and M. C. Thurber (eds), *Oil and Governance: State-Owned Enterprises and the World Energy Supply*, Cambridge: Cambridge University Press.

Vivoda, V. (2009) 'Resource Nationalism, Bargaining and International Oil Companies: Challenges and Change in the New Millennium', *New Political Economy*, 14(4): 517–34.

Volman, D. (2003) 'The Bush administration and African Oil: The Security Implications of US Energy Policy', *Review of African Political Economy*, 30(98): 573–84.

Von der Mehden, F. R. and Troner, A. (2007) 'Petronas: A National Oil Company with an International Vision', *Policy Brief*, The James A. Baker Institute for Public Policy, Rice University.

Wagner, J. and Armstrong, K. (2010) 'Managing Environmental and Social Risks in International Oil and Gas Projects: Perspectives on Compliance', *Journal of World Energy Law and Business*, 3(2): 140–65.

Walker, R. A. (2001) 'California's Golden Road to Riches: Natural Resources and Regional Capitalism', *Annals of the Association of American Geographers*, 91: 167–99.

Wallerstein, I. (1974) *The Modern World-System: Capitalist Agriculture and the Origins of the European World-Economy in the Sixteenth Century*, Berkley, CA: University of California Press.

Walt, S. (1987) *The Origins of Alliances*, Ithaca, NY: Cornell University Press.

Waltz, K. N. (1979) *Theory of International Politics*, New York: McGraw-Hill Higher Education.

Ward, B. and Dubos, R. (1973) *Only One Earth*, London: Penguin Books.

Warshaw, C. (2012) 'The Political Economy of Expropriation and Privatization in the Oil Sector', in D. G. Victor, D. R. Hults and M. C. Thurber (eds), *Oil and Governance: State-Owned Enterprises and the World Energy Supply*, Cambridge: Cambridge University Press.

Watkins, M. H. (1963) 'A Staple Theory of Economic Growth', *Canadian Journal of Economics and Political Science*, 29: 141–158.

Watson, H. J. and Bankes, N. (2010) 'Different Views of the Cathedral: The Literature on Property Law Theory', in A. McHarg, B. Barton, A. Bradbrook and L. Godden (eds), *Property and the Law in Energy and Natural Resources*, Oxford: Oxford University Press.

Watson, J. (1997) 'The Technology that Drove the "Dash for Gas" ', *Power Engineering Journal*, 11(1): 11–19.

Watts, M. (2001) 'Petro-Violence: Community, Extraction and Political Ecology of a Mythic Commodity', in N. L. Peluso and M. Watts (eds), *Violent Environments*, Ithaca, NY: Cornell University Press.

Watts, M. (2004a) 'Resource Curse?: Governmentality, Oil and Power in the Niger Delta, Nigeria', *Geopolitics*, 9(1): 50–80.

Watts, M. (2004b) 'Violent Environment: Petroleum Conflict and the Political Ecology of Rule in the Niger Delta', in R. Peet and M. Watts (eds), *Liberation Ecologies (Environment, Development, Social Movements)*, 2nd edn, London: Routledge.

Watts, M. (ed.) (2008) *Curse of the Black Gold: 50 Years of Oil in the Niger Delta*, New York: Powerhouse Books.

Watts, M. (2009) *Crude Politics: Life and Death on the Nigerian Oil Fields*, Berkeley, CA: University of California, Institute of International Studies.

WB (World Bank) and IFC (International Finance Corporation) (2002) *Large Mines and Local Communities: Forging Partnerships, Building Sustainability*, Washington, DC: International Finance Corporation.

Weaver, C. and Gunton, T. (1982) 'From Drought Assistance to Megaprojects: Fifty Years of Regional Theory and Policy in Canada', *The Canadian Journal of Regional Science*, 1: 5–37.

Weber-Fahr, M., Strongman, J., Kunanayagam, R., McMahon, G. and Sheldon, C. (2001) *Mining and Poverty Reduction*. Available at http://bit.ly/jFV8fi [Accessed 25 May 2011].

Wells, L. T. and Ahmed, R. (2007) *Making Foreign Investment Safe: Property Rights and National Sovereignty*, Oxford: Oxford University Press.

Wells, P. R. A. (2005a) 'Oil Supply Challenges-1: The Non-OPEC Decline', *Oil & Gas Journal*, 21 February.

Wells, P. R. A. (2005b) 'Oil Supply Challenges-2: What can OPEC Deliver?', *Oil & Gas Journal*, 7 March.

Wendt, A. (1992) 'Anarchy is What States Make of It: The Social Construction of Power Politics', *International Organization*, 41(3): 391–425.

Wenger, A., Orttung, R. W. and Perovic, J. (eds) (2009) *Energy and the Transformation of International Relations: Toward a New Producer-Consumer Framework*, Oxford: Oxford University Press, Oxford Institute for Energy Studies.

Weyant, J. P., de la Chesnaye, F. C. and Blanford, G. J. (2006) 'Overview of EMF-21: Multigas Mitigation and Climate Policy', *The Energy Journal*, Multi-Greenhouse Gas Mitigation and Climate Policy (Special Issue #3): 1–32.

Whitmore, A. (2006) 'The Emperors New Clothes: Sustainable Mining?' *Journal of Cleaner Production*, 14: 309–314.

Williamson, O. E. (2000) 'Economic Institutions and Development: A View from the Bottom', in M. Olson and S. Kähkönen (eds), *A Not-so-Dismal Science: A Broader View of Economics and Society*, Oxford: Oxford University Press.

Wohlforth, W. C. (1999) 'The Stability of a Unipolar World', *International Security*, 24(1): 5–41.

Wood Mackenzie (2010) *The Growth of Asian NOCs in the Middle East*, Wood Mackenzie.

Wood, T. (2007) 'Contours of the Putin Era', *New Left Review*, 44: 53–71.

World Bank (2008) *Mining and Poverty Reduction: Oil, Gas, Mining and Chemicals: Key Topics in Mining*, Washington, DC: The World Bank.

Woynillowicz, D., Severson-Baker, C. and Raynolds, M. (2005) *Oil Sands Fever: The Environmental Implications of Canada's Oil Sands Rush*. Available at Pembina Institute: pubs.pembina.org/reports/OilSands72.pdf [Accessed 20 August 2011].

Wright, G. and Czelusta, J. (2007) 'Resource-Based Growth: Past and Present', in D. Lederman and W. F. Maloney (eds), *Natural Resources: Neither Curse Nor Destiny*, Washington, DC: World Bank and Stanford University Press.

Xu, X. (2007) 'Chinese NOC's Overseas Strategies: Background, Comparisons and Remarks', *Policy Brief*, The James A. Baker Institute for Public Policy, Rice University.

Yafimara, K. (2011) *The Transit Dimension of EU Energy Security: Russian Gas Transit across Ukraine, Belarus and Moldova*, Oxford: Oxford University Press.

Yates, D. (1996) *The Rentier State in Africa: Oil Rent Dependency and Neocolonialism in the Republic of Gabon*, Trenton: Africa World Press.

Yergin, D. (1991) *The Prize: The Epic Quest for Oil, Money and Power*, New York: Simon and Schuster.

Yergin, D. (2011) *The Quest: Energy, Security, and the Remaking of the Modern World*, London: Penguin Books.

Yessenova, S. (2007) 'Tengiz Crude: A View from Below', in B. Najman, R. Pomfret and G. Raballand (eds), *The Economics and Politics of Oil in the Caspian Basin: The Redistribution of Oil Revenues in Azerbaijan and Central Asia*, London: Routledge.

Zakaria, F. (2008) *The Post-American World and the Rise of the Rest*, New York: W. W. Norton.

Zalik, A. (2004) 'The Niger Delta: Petro Violence and Partnership Development', *Review of African Political Economy*, 31(101): 401–24.

Zanoyan, V. (2007) 'State Owned and Private Multinational Enterprise – a Paradigm Shift (pre-seminar reading)', *Global Leadership Seminar*, Tufts Centre: Talloires.

Zhao, L., Lianyong, F. and Hall, Ch. A. S. (2009) 'Is Peakoilism Coming?' *Energy Policy*, 3(6): 2136–8.

Index

Note: Locators followed by *f* and *t* refer to figures and tables.

Bush, George W., 29
business relationships, 53

Campbell, Colin, 187, 190
camping gas, 61
cannibalism, 130
capitalism, conceptualization of, 7
carbon tax, 202
cartel, 20, 42, 92, 124–7, 139, 193,
    233–4
Carter doctrine, 82
Chad-Cameroon pipeline project, 111
Chile Copper Company, 35
China
  'China threat' debate, 96
  commodity price boom, 49
  conflicts with US, 83
  demand for oil and gas, 136
  downturn in commodity prices, 50
  economic growth, 49
  economic liberation, 50
  global mineral share, 49f
  'go out' policy, 50–1
  impact on international development,
    95
  imposition of limits on export, 56
  industrial giants, 57
  integration into liberal international
    system, 95
  LNG import, 71
  no preferential treatment, 136
  presence in Africa and Latin America,
    84
  principle of non-interference, 51
  quest for mineral supplies, 51
  rare earth metals, 56
  rise as emerging powers, 232
  shift of power, 95
  South China Seas, sovereign
    claims, 84
  state capitalism, 112
China Investment Corporation (CIC), 50
China Minmetals, 50
China Nonferrous Metals Corporation
    (CNMC), 50–1
Chinese National Offshore Oil
    Corporation (CNOOC), 129
Churchill, Winston, 121
classical economy, 99
Clean Water Act, 217

Cleveland LNG disaster (1944), 68
climate change, 5, 8, 27, 29, 57, 59, 69,
    137–8, 140, 155, 201, 203
club goods hypothesis, 180
Club of Rome, 56
clubs of beneficiaries, 180
coal-to-liquid, 30, 204
Coase theorem, 174–6, 178
Cold War, 36–9
  bi-polarity of, 54
  copper price since 1950, 38f
  demand for minerals growing, 37
  influence of, 39–40
  international politics of, 82
  notion of nationalization, 37
  Second World War, effect, 36
  US geopolitical hegemony, 36
Collier, Paul, 108–9
Colombia, state capitalism, 104
colonialism, 34
colonial legacies, 95
combined-cycle gas turbine power
    station (CCGT), 67
commodity prices, decline in, 147, 232
community conflicts, 44–5
  community engagement to foster
    cooperation, 45
  fault lines of conflict, 44
  human rights, 45
  political divisions, 45
  religious divisions, 45
companies regime and state support,
    123–5
  conflict with host and parent state,
    124–5
  economic miracle, 124
  equity participation, 123
  post-war reconstruction and
    prosperity, 124
  sisters' regime, 123–4
  stunt competition, 124
  three 'sisters', 123
company-communities relations, 234
comparative advantage theory, 99, 102
competition, 125–8
  bargaining position, 126
  new entrants (independents), 126
  sisters regime, 126–7
Congo (Shaba province), 41
congressional report, 214